TUNNELS:
Planning, Design, Construction

Vol. 2

ELLIS HORWOOD SERIES IN ENGINEERING SCIENCE

STRENGTH OF MATERIALS
J. M. ALEXANDER, University College of Swansea.
TECHNOLOGY OF ENGINEERING MANUFACTURE
J. M. ALEXANDER, R. C. BREWER, Imperial College of Science and Technology, University of London, J. R. CROOKALL, Cranfield Institute of Technology.
VIBRATION ANALYSIS AND CONTROL SYSTEM DYNAMICS
CHRISTOPHER BEARDS, Imperial College of Science and Technology, University of London.
COMPUTER AIDED DESIGN AND MANUFACTURE
C. B. BESANT, Imperial College of Science and Technology, University of London.
STRUCTURAL DESIGN AND SAFETY
D. I. BLOCKLEY, University of Bristol.
BASIC LUBRICATION THEORY 3rd Edition
ALASTAIR CAMERON, Imperial College of Science and Technology, University of London.
STRUCTURAL MODELLING AND OPTIMIZATION
D. G. CARMICHAEL, University of Western Australia
ADVANCED MECHANICS OF MATERIALS 2nd Edition
Sir HUGH FORD, F.R.S., Imperial College of Science and Technology, University of London and J. M. ALEXANDER, University College of Swansea.
ELASTICITY AND PLASTICITY IN ENGINEERING
Sir HUGH FORD, F.R.S. and R. T. FENNER, Imperial College of Science and Technology, University of London.
INTRODUCTION TO LOADBEARING BRICKWORK
A. W. HENDRY, B. A. SINHA and S. R. DAVIES, University of Edinburgh
ANALYSIS AND DESIGN OF CONNECTIONS BETWEEN STRUCTURAL JOINTS
M. HOLMES and L. H. MARTIN, University of Aston in Birmingham
TECHNIQUES OF FINITE ELEMENTS
BRUCE M. IRONS, University of Calgary, and S. AHMAD, Bangladesh University of Engineering and Technology, Dacca.
FINITE ELEMENT PRIMER
BRUCE IRONS and N. SHRIVE, University of Calgary
PROBABILITY FOR ENGINEERING DECISIONS: A Bayesian Approach
I. J. JORDAAN, University of Calgary
STRUCTURAL DESIGN OF CABLE-SUSPENDED ROOFS
L. KOLLAR, City Planning Office, Budapest and K. SZABO, Budapest Technical University.
CONTROL OF FLUID POWER, 2nd Edition
D. McCLOY, The Northern Ireland Polytechnic and H. R. MARTIN, University of Waterloo, Ontario, Canada.
TUNNELS: Planning, Design, Construction — Vols. 1 and 2
T. M. MEGAW and JOHN BARTLETT, Mott, Hay and Anderson, International Consulting Engineers
UNSTEADY FLUID FLOW
R. PARKER, University College, Swansea
DYNAMICS OF MECHANICAL SYSTEMS 2nd Edition
J. M. PRENTIS, University of Cambridge.
ENERGY METHODS IN VIBRATION ANALYSIS
T. H. RICHARDS, University of Aston, Birmingham.
ENERGY METHODS IN STRESS ANALYSIS: With an Introduction to Finite Element Techniques
T. H. RICHARDS, University of Aston, Birmingham.
ROBOTICS AND TELECHIRICS
M. W. THRING, Queen Mary College, University of London
STRESS ANALYSIS OF POLYMERS 2nd Edition
J. G. WILLIAMS, Imperial College of Science and Technology, University of London.

TUNNELS:
Planning, Design, Construction

Vol. 2

T. M. MEGAW, M.Sc., Hon.D.Sc., F.I.C.E.
Consultant and Former Senior Partner

and

J. V. BARTLETT, C.B.E., M.A., F.Eng., F.I.C.E., F.I.E.Aust., F.A.S.C.E.
Director and Senior Partner
Mott, Hay & Anderson, Consulting Civil Engineers
Croydon, Surrey

ELLIS HORWOOD LIMITED
Publishers · Chichester

Halsted Press: a division of
JOHN WILEY & SONS
New York · Chichester · Brisbane · Toronto

First published in 1982 by

ELLIS HORWOOD LIMITED
Market Cross House, Cooper Street, Chichester, West Sussex, PO19 1EB, England

The publisher's colophon is reproduced from James Gillison's drawing of the ancient Market Cross, Chichester.

Distributors:

Australia, New Zealand, South-east Asia:
Jacaranda-Wiley Ltd., Jacaranda Press,
JOHN WILEY & SONS INC.,
G.P.O. Box 859, Brisbane, Queensland 40001, Australia

Canada:
JOHN WILEY & SONS CANADA LIMITED
22 Worcester Road, Rexdale, Ontario, Canada.

Europe, Africa:
JOHN WILEY & SONS LIMITED
Baffins Lane, Chichester, West Sussex, England.

North and South America and the rest of the world:
Halsted Press: a division of
JOHN WILEY & SONS
605 Third Avenue, New York, N.Y. 10016, U.S.A.

© 1982 T. M. Megaw and J. V. Bartlett/Ellis Horwood Ltd.

British Library Cataloguing in Publication Data
Megaw, T. M.
Tunnels. – (Ellis Horwood series in engineering science: mechnical and civil engineering)
Vol. 2
1. Tunnels
I. Title   II. Bartlett, J. V.
624.1'93   TA805

Library of Congress Card No. 81–4111 AACR2

ISBN 0–85312–361–6 (Ellis Horwood Ltd., Publishers)
ISBN 0–470–27209–0 (Halsted Press)

Typeset in Press Roman by Ellis Horwood Ltd.
Printed in Great Britain by R. J. Acford, Chichester.

Table of Contents

Authors' Preface

The authors covered in Volume 1 the general field of tunnelling, more particularly bored tunnels, and they hope that is a worthwhile book in its own right. The Preface to Volume 1 outlined the general contents of Volume 2 as follows: tunnels in trench under land and water, shafts, ground treatment, geological and geotechnical aspects, and specialised applications for road and rail transport systems. Inevitably, it has been necessary to limit severely the discussion of every subject treated, but it is hoped that the chapter bibliographies will provide a useful lead deeper into the various subjects.

Again the emphasis has been on describing methods and principles and indicating choice to be made rather than on development and application of theoretical analysis in which engineering geology, soil mechanics and rock mechanics have so much to contribute. The aim is not to give detailed instruction on any subject, but to show what factors have to be considered in the perspective of the whole complex art.

An extensive bibliography of publications in English has been included, of wider scope than the chapter bibliographies, although incidentally including most of their contents to make it comprehensive. A vast amount of information is to be found in the publications of the Institution of Civil Engineers and of the American Society of Civil Engineers. These each have been made the subject of separate lists, in order of date of publication, and chronological order has similarly been followed in the general list of books and articles. Ever more numerous specialist conferences are occasions of presenting papers, some of major importance, some of local and temporary interest only. The conferences have been listed with an indication where possible of any particular theme relevant to this book. Individual papers have been included in the general bibliography if thought appropriate but many papers of great value are not separately identified.

In addition to the bibliography an Index of Tunnels named in the book (Volumes 1 and 2) is appended, noting their locations and functions.

New developments

New techniques and equipment are being introduced so rapidly that no attempt
to describe and evaluate them has been possible. A very few may be mentioned
here, of which detailed descriptions must be sought in the periodicals and in
conference and other papers.

The recent successful use of an earth pressure balance tunnelling machine by
a Japanese contractor in San Francisco is just one of many indications of the
rapid changes in the tunnelling world. Two achievements of particular note on
that sewer tunnel project were that the cutting tools of the TBM lasted throughout the drives without replacement, and secondly a number of timber piles were
successfully 'chewed through' by the cutting head in the course of the tunnelling.
That opens up new possibilities of shallow bored tunnelling in many other cities
(in the Low Countries of Europe for instance) provided of course that the piles
are no longer in important use as supports.

A recent promising innovation is the British 'Unitunnel' system. Major
disadvantages of long distance pipe jacking are the high jacking loads and steering
difficulties. The Unitunnel system overcomes both difficulties by jacking the
pipe sections (or tunnel rings) forward one at a time. That is achieved by successively inflating and exhausting pneumatic hoses between each pipe (or ring) in
such a way that one pipe at a time is being pushed forward off the two pipes
behind it, and so on in turn until an additional pipe can be inserted in the
working shaft or pit at the start of the drive: for most proposed applications,
a simple annular cutting edge with a few steering jacks is all that will be required
in the way of a shield, but excavation and face support can be organised in a
variety of ways.

Meanwhile research continues on the use of novel forces such as laser beams
and high pressure water jets for the excavation of rock, the latter having reached
the development stages for softer rocks.

The most intractable tunnelling problem remains our inability to picture
exactly the ground and water conditions which will be encountered during the
drive. For some tunnels one of the most helpful developments is the new ability
to navigate a probe for considerable distances ahead of the working face. But
the expense is considerable and much uncertainty of ground and water conditions
is likely to remain to challenge the tunneller.

Whilst tunnels will progressively be constructed more economically and
safely than in the past, the authors see no danger of this branch of civil engineering losing its fascination for those who enjoy dealing occasionally with the
unusual and the unexpected.

Acknowledgements

We reiterate our gratitude to those who have contributed to the text and the
production of the book, whether or not named in the previous preface. Illus-
trations from some new sources appear in this volume: we express our thanks to
all contributing.

1

Cut and Cover

1.1 CHARACTERISTICS

Construction of a tunnel by cut-and-cover offers an alternative to boring where a trench of the required depth and width can be excavated from the surface. In its simplest form, a trench is excavated, the tunnel structure is built, the trench is backfilled and the surface is restored, but the support of soft ground and the maintenance of existing surface and underground facilities and services make most projects much more complex.

For shallow tunnels the direct cost of cut-and-cover is likely to be much less than the cost of boring, but the incidental costs can change the balance completely. These include such costs, not always ascertainable or necessarily chargeable, as provision of alternative facilities for traffic using the surface, safeguards against subsidence, protection or diversion of services and drainage systems and social costs of disruption and loss of amenity.

With increasing depth, direct costs of trench excavation and support increase rapidly. In water-bearing ground the water must be managed by containment, pumping or groundwater lowering, while, in soft clays particularly, heave of the trench bottom may lead to serious loss of ground and subsidence.

For metro tunnels in particular there is unending debate on the choice between bored tunnels and cut-and-cover. Direct construction costs and operating economics obviously favour shallow cut-and-cover, but very shallow tunnels are rarely practicable and all the incidental costs and benefits must be taken into consideration. There is no universal answer: each city, and indeed each line in that city, has its own special problems.

The first London underground lines were built as cut-and-cover but objections to disruption, and incidental costs made it impossible to obtain the necessary powers to construct except by deeper bored tunnelling, adopted for all later lines in the central area. An interesting recent development has been the extension to Heathrow Airport, partly in cut-and-cover and partly, within the airport perimeter, in bored tunnel. Toronto presents another example of a

city where the first line was built in cut-and-cover, but for subsequent work in
the city extensive shield tunnelling, some of it with compressed air, has been
adopted.

1.2 AREAS OF USE

Cut-and-cover tunnels are characteristically shallow tunnels in situations where
open cuttings are not acceptable in the finished project. They are most commonly
employed:

1. in subaqueous tunnels to form a transition between an
   open cut approach and the main tunnel;
2. at the portals of mountain tunnels;
3. in urban conditions where the route must be covered
   in and the surface restored; and
4. for reasons of amenity.

1.2.1 Subaqueous tunnels

In subaqueous tunnels a transition may connect an open approach ramp with a
shield bored or submerged unit tunnel. At the outer end the preference for a
covered tunnel over open cut may arise from:

(a) construction problems and costs for open cut;
(b) need to restore surface use over the tunnel; and
(c) danger of flooding.

(a) In the vicinity of a river the ground is usually alluvial such as soft silts
and clays, water-bearing gravels with perhaps peaty layers. Any trench excavation
will require substantial lateral support and management of water with precautions
against bottom heave. While temporary support for a working trench may be
feasible, permanent retaining walls and their foundations may become unduly
heavy and costly and require such top strutting that a closed box structure is
preferable. Upward water pressure acting on the base of an open cut may have
to be counteracted by downward loading, more economically provided by backfill over a tunnel roof.

(b) In open country a road or railway close to the river bank may pass
across the tunnel approach at ground level or a flood bank may have to be
restored. Drainage systems, whether in open channels or pipes, might be
obstructed by a cutting. In urban conditions, the surface requirements are
as more fully discussed below.

(c) Protection against flooding by abnormal tides may be more satisfactorily
given by roofing over than by high flood banks or walls. Where construction is
in a cofferdam extending into the waterway the need for a roofed structure is
obvious.

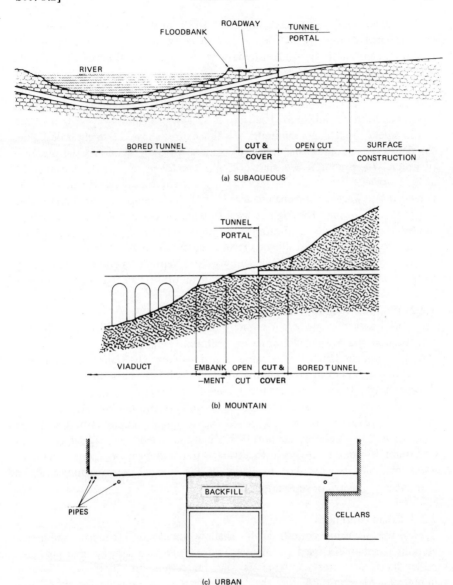

Fig. 1.1 – Principal uses of cut and cover: (a) Subaqueous, (b) Mountain, (c) Urban.

At the deep end of the transition the appropriate point of change may be determined by physical limitations defining a minimum depth for the main tunnel technique or a maximum for the cut-and-cover excavation. Comparative costs and access for construction may also be decisive.

For submerged unit tunnels sufficient depth for flotation is critical, but that is determined by the practicability and economy of dredging. An interesting case was in an early metro line in Amsterdam where a canal was dredged along a wide approach highway allowing units to be floated in, sunk and joined. It is understood that the scale of excavation and disruption proved to be unacceptable and the method has not been used subsequently. The maintenance of essential services and utilities must have been another major problem.

In shield tunnelling a minimum of a few metres above the shield is desirable, the depth depending on the cohesive character of the ground and the presence of water. Where compressed air working is employed the overburden must be at least sufficient to retain the necessary air pressure, although the drive can be extended landwards on a rising gradient by building a temporary bank to provide the required loading. The highest level to which a shield is driven may also be limited by the need for a stratum at the invert sufficiently firm to support the weight of the shield and to allow its level and attitude to be controlled.

It is frequently found convenient to site a ventilation station, possibly by caisson construction, at the transition from main tunnel to cut-and-cover.

1.2.2 Mountain tunnels

The portal area of a mountain tunnel presents considerable difficulties both in construction and in the final structure. Failure to appreciate the need for special care at the portal has been the cause of serious falls. The transition from open cut to bored tunnel in competent rock must normally pass through a surface slope of unstable scree and detritus underlain by weathered rock, and in many cases cut-and-cover construction is the most efficient method. It allows the ground above the portal to be exposed and adequately explored before tunnel excavation and overhead support are finally decided, and subsequent roof construction can be designed to divert water and protect against rockfalls. There may also be advantage in extending the covered construction outwards if avalanches of snow or scree are to be feared.

1.2.3 Urban conditions

Typical uses, in urban conditions, for shallow tunnels are for metros and metro stations (further discussed in Chapter 7), for pedestrian subways and pipe and cable ducts, for street underpasses and for sewers. In city centres, and in diminishing degree outwards, the land surface is so valuable for other uses that an open cut is not practicable. The choice is not between open cut and tunnel, but between cut-and-cover and bored tunnel construction. Minimum depth, and therefore normally cut-and-cover, is essential for pedestrian subways or street underpasses if they are to perform their functions. Throughout the history of metros the functional ideal has been to site them at the shallowest depth possible so that access between street and platform is quick and easy. The immediate obvious answer is cut-and-cover. In practice, bored tunnelling may be found necessary because street disruption, resulting from surface excavation, is unacceptable, or because existing underground obstacles, such as buried services, sewers or other tunnels crossing the line, force down the tunnel to depths where cut-and-cover ceases to be economic. As a city elaborates its underground network this factor of routes crossing one another becomes of increasing importance.

Fig. 1.2 – Metro construction by cut and cover in Rotterdam. Disruption to surface traffic is obvious.

Where the route is not so obstructed many ingenious methods have been adopted to minimise the time and extent of the disruption resulting from open excavation.

Even where metro running tunnels are at a depth where boring is necessary the complex station structures may, wholly or partly, be built in deep open pits. In some cases an 'umbrella' has first been built to carry the street traffic while excavation and construction proceed below.

Sewer systems range from small pipes laid in trench at the upper end of the system to progressively greater diameters and greater depths at which tunnelling may become necessary, either because of depth or because trench excavation occupying the surface is unacceptable in busy streets or under buildings. Even for small pipes excavation in heading may be necessary.

Fig. 1.3 – Metro 'umbrella'. (a) Trial assembly of material for use at crossroads above London Fleet Line station. (b) 'Umbrella' in place.

1.2.3.1 Street work

The essentials of cut-and-cover construction are: excavation of trench; building of floor, walls and roof; backfilling and restoration of surface. Particularly in urban conditions, complex procedures are usually found necessary. If the line of a street is being followed the principal disadvantages in construction special to this method, as compared to a deeper bored tunnel, are:

  (a) Lengthy occupation of street with consequent noise and disturbance and disruption of traffic and access;

  (b) restricted alignment;

  (c) necessity of lateral support to minimise subsidence;

  (d) interference with buried services—water, gas, sewers, etc.—which must be diverted or supported; and

  (e) backfill material and restoration.

  (a) The disturbance caused by opening up long lengths of a street can be greatly reduced, but never eliminated, by alternating excavation and construction for the separate elements of the tunnel. The side walls can be constructed first, occupying only a relatively narrow part of the street, the roof structure can then be built in short lengths which require only shallow excavation, allowing the street surface to be restored quickly, and providing strutting at the top of the walls. The excavation of the 'dumpling' can follow as in a bored tunnel without surface occupation or disturbance. The floor excavation may have to be done with particular care in narrow widths if there may be ground heave or if the side walls require to be strutted at the base.

  Such a method does not require lengthy occupation of the whole street surface, but does still imply extensive obstruction of traffic as long lengths of wall construction obstruct lateral access and the whole width of the structure across the street must be occupied as each length of roof is constructed. In suitable conditions or for special lengths, it may be possible to start construction with the roof slab and to build the walls in tunnelled headings to underpin the edges of the roof slab.

  (b) For metro and railway work in particular the need to follow an alignment within the limits of existing streets can be a serious disadvantage. Small radius curves may be unavoidable and the direct route between preferred sites for stations may not be possible. Basements and cellars extending beyond the building line impose further limitations.

  (c) Lateral support. Open excavation with the sides at the natural slope of the ground is unlikely to be possible in street work, except in rock. Vertical sides reduce the width occupied to a minimum, but inevitably require lateral support, which may either be temporary or may embody the permanent wall structure.

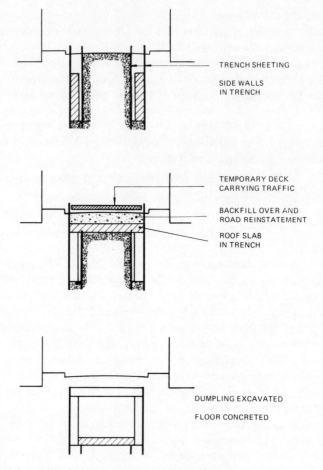

Fig. 1.4 – Cut and cover in street – normal sequence.

Although the tunnels under consideration are described as 'shallow', the trenches and pits required are deep by ordinary excavation standards and some discussion of settlement, and methods of keeping it to a minimum is required.

In any deep trench through cohesive soils the differential pressure on the side walls of the trench will result in strain displacement in the retained soil. There will inevitably be a tendency to plastic heave at the bottom of the trench and inward deflection of the retaining wall with resultant loss of ground and settlement at the surface. The magnitude of this depends on the nature of the soil, the depth of excavation, and the characteristics of the support system. Its importance will depend on the nearness and depth of foundations of adjacent structures and their sensitivity to movement. Strain movements can be minimised by strutting apart the sides at frequent vertical intervals as excavation proceeds. The prestressing of ground anchors and the incorporation of jacking devices in struts, and structurally stiff wall sections can probably reduce the deflections to acceptable proportions.

Ground water is of major importance. If it is retained by watertight piling, the water pressure must be carried by the lateral support system, and unless there is an effective cut off the full upward pressure will act on the bottom. If the water table is lowered by pumping, well pointing or otherwise there may be danger of consequent settlement in adjacent shallow foundations. Old buildings, supported on timber piles can be very sensitive to such changes in water level.

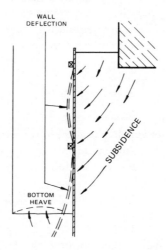

Fig. 1.5 – Pattern of subsidence adjacent to trench.

(d) Buried services are a major problem in cities where they have been laid —and not always accurately recorded—under the street surface at varying depths. Water and gas pipes up to 1½ m diameter are not uncommon; sewers may range from 100 mm house connections to trunk sewers of 2 m diameter upwards; high voltage cables, in ducts or armoured, and banks of telephone cables in ducts are likely to be encountered. Tramway conduits, hydraulic power pipes, and pneumatic despatch tubes are other hazards that may be present in some cities.

Where a cut-and-cover tunnel is to be built these services must be fully safeguarded during and after construction. Diversion beforehand out of the area of excavation is ideal for the tunnel work but is rarely possible for all of the services, although temporary diversion, or even cutting off, is sometimes practicable. Otherwise they have to be supported, protected and maintained across the open excavation and in due course must be very carefully reinstated and secured against subsidence. While so supported they constitute a continuing obstacle to excavation and to supports for any temporary decking. A fractured or leaky gas or water main can be a serious hazard in an enclosed excavation.

(e) Backfill. Restoration of the surface above the roof to its original use will require careful backfilling to avoid subsidence and to ensure that natural drainage is preserved.

In streets, as already noted, backfilling under and around buried services will require particular care. Reconstruction of the road slab and its surfacing and foundations will be necessary.

At greater depth it may be possible to utilise the excavated material as backfill, but unless it is suitable new material must be imported.

In parks and open spaces the proper restoration will demand careful attention to drainage and replacement of topsoil with adequate but not excessive compaction. Safeguarding of trees during construction and restoration may also be called for.

1.2.4 Amenity

A further reason, of growing importance, for enclosing traffic routes is that of local amenity in order to protect residential areas, parks, scenic areas and nature reserves against visual and acoustic disturbance, and interference with existing surface pathways and access.

Motorways in scenic areas or leading into cities across green belt land may be thus built below ground and roofed over. In by-pass construction skirting historic towns or villages, similar protection of amenity may be required. For such cases shallow cut-and-cover construction rather than boring is usually appropriate.

1.3 TECHNIQUES OF CONSTRUCTION

The key to the various methods lies largely in the support of the vertical sides. The principal techniques include:

1. Steel sheet piling with walings and struts or ground anchors.
2. Ground anchors.
3. King piles with struts or ground anchors and horizontal poling boards.
4. Slurry trench walls.
5. Large diameter bored piles contiguous or overlapping.
6. Concrete walls in heading.

The choice of method will be determined by the nature of the ground, the assessment of the difficulties listed above, available resources, cost of construction, land costs and social costs.

1.3.1 Steel sheet piling

This is the simplest engineering solution in soft ground provided it is free from obstructions natural or artificial, such as boulders or cellars and foundations or service pipes leading to and from buildings: the piling requires to be supported as excavation proceeds by walings and struts forming a framework wedged into place. Struts across the trench make access for excavation difficult: ground anchors are a possible alternative. The vertical intervals between frames are calculated initially to give acceptable stresses and deflections in the piles, but even when set and wedged tightly as early as possible some deformation of the ground will occur as noted above, because of ground stresses and plastic flow as the trench is deepened. Movement can be reduced by imposed loading in struts or prestressing in anchor cables, but the flexibility of the steel sheet piling makes unavoidable some loss of ground and resultant settlement, which may or may not be within acceptable limits at the distance from the piling of the nearest structure. Another possible source of settlement is the compaction of loose granular soil resulting from vibration.

Pile driving by hammer is a noisy operation but relatively quiet methods have been developed. Headroom for pitching a line of piles may not be available, particularly in city streets, where there are overhead wires, or where close to buildings.

Steel sheet piling provides an effective temporary barrier to water except where piles get distorted and out of clutch. In water-bearing ground this may necessitate remedial measures such as local ground injection treatment or, possibly, additional piles. Watertightness of the piling is rarely adequate for permanent tunnel requirements. The piling must frequently be left in place although the tops may be burned off at some stage of excavation and backfill.

1.3.2 Ground anchors

One great merit of ground anchors is that they avoid any obstruction of the trench by struts. Another is that they can be readily prestressed. They do, however, extend usually far behind the line of the wall, and very probably beyond the limits of acquisition of land. It will then be necessary first to verify that the anchorages are clear of basements or other structures, and then to obtain permission from the adjacent property owners. For tunnel construction the use of ground anchors will almost certainly be temporary, until the tunnel roof is in place and acts as a strut. If, for any reason, permanent anchorages are required, they would have to be fully protected, not only against corrosion, but against disturbance or damage caused by future site development.

USE OF GROUND ANCHORS

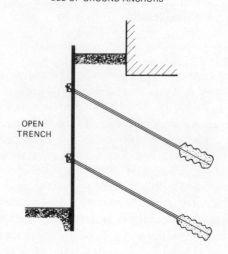

OPEN
TRENCH

Fig. 1.6 – Section of sheet piling with walings and ground anchors.

1.3.3 King piles

An alternative is the use of bored king piles at regular spacing. They may be of heavy section steel lowered into a pre-bored hole. Where rock is present at suitable depth the toe can be grouted into the rock. The piles will require strutting across the trench as excavation proceeds, but ground anchors, pre-stressed if necessary, will minimise obstruction of the area of excavation. The

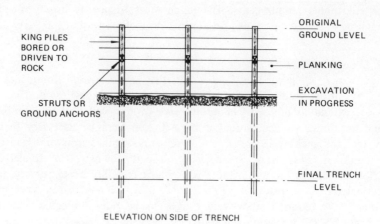

ELEVATION ON SIDE OF TRENCH

Fig. 1.7 – King piles.

Fig. 1.8 – Open cut approach to Mersey Kingsway Tunnel, supported by king piles and ground anchors.

ground between the piles is supported by horizontal poling boards set in place
as excavation proceeds. A variant is where a length of wall is concreted in a
slurry filled trench excavated between the king piles. The piles can be designed
to carry a temporary road deck, beneath which excavation without further
disturbance to traffic is possible.

As with steel sheet piling, possible settlement of the adjacent ground must
be considered. Headroom, first for boring, and then for setting the piles, has to
be ensured. Obstacles can be dealt with at the boring stage.

1.3.4 Slurry trench walls

Concrete walls cast in slurry filled trenches are a comparatively recent development. A trench is excavated without lateral support except that provided by
bentonite or other dense clay slurry filling the trench throughout the process.

Fig. 1.9 – Cut and cover approach to second Dartford Tunnel: slurry trench walls.

Fig. 1.10 – Use of concrete bored piles in Liverpool approach to Mersey Kingsway Tunnel. (a) Contiguous pile walls completed and tops exposed. Arch shutter in place. Walls above with king piles and ground anchors. (b) Structure of cut and cover completed. *Note*: exposed shafts of bored piles (subsequently covered), *in situ* arch over, two main tunnel portals (11 m dia.) in headwall at rear.

Within this trench a concrete wall can be cast by tremie, with a reinforcement cage if required. The concrete wall so cast may either be temporary, performing the same function as sheet piling, or may form or be embodied in the permanent structure.

In cut-and-cover tunnelling this has the advantage of minimising surface disruption and obstruction, and of requiring less headroom than sheet piling, although some headgear is necessary. Also, the whole operation is quieter. The permanent wall can be thus constructed, but the finish, if cast against the ground, will be rough and will normally require to be faced up. This wall will be substantially more rigid than sheet piling, although struts or anchors will still be required during excavation of the main trench. Some surface settlement is inevitable but it can be reduced to a very small magnitude. A hazard with slurry trenching is that if unexpected voids are encountered the slurry may be lost and the trench sides may collapse. Possible voids in urban areas include unmapped basements, abandoned and fractured pipes and sewers, or old wells.

1.3.5 Concrete bored piles
Large diameter bored piles at close centres concreted *in situ* can be used to form a wall. They may be contiguous or may overlap ('secant' piling). The piles may be taken down to any required foundation level and can be reinforced. In the case of secant piling alternate piles, driven first, may be left unreinforced because they will be sliced down in boring the intermediate piles. With good workmanship substantial watertightness is possible, particularly so with secant piling. The concrete at the head of the piles can be finished to take the roof structure covering the tunnel.

In the Mersey Kingsway tunnel at the Liverpool approach the twin bored tunnels converge, where the depth to roadway is 18 m, to a single cut-and-cover tunnel 31.5 to 26 m wide at road level and about 244 m long. Contiguous bored piles, 2.4 m in diameter and keyed into rock, form the walls, and are strutted apart at the top by an arched roof with backfill over. In this instance the principal difficulty of open cut was that cantilevered retaining walls would have been excessively massive, and struts spanning across the top would have been long and heavy and exposed to temperature stresses. Apart from one street crossing the line of the tunnel, the land above was not built over.

1.3.6 Walls in heading
In special cases where access from the surface is particularly difficult, concrete side walls can be built in a heading, either so arranged that the roof slab will be seated on them subsequently, or to underpin a roof slab already in position.

BIBLIOGRAPHY

Cut and cover in general as a form of tunnel construction does not appear to be the subject of much separate writing. It is usually described and discussed as incidental to particular projects, especially metro systems and urban underpasses.

In Paris and Brussels a large number of street underpasses were built in recent years and are described in the P.I.A.R.C. documents noted below.

The following brief list includes general papers and descriptions of a few works; more appear in the general bibliography. Reference should also be made to Chapter 7 — Metro Tunnels.

Parsons, W. B., The New York rapid-transit subway, *Min. Proc. Instn Civ. Engrs*, 1908, **173**.

Humphreys, G. W., The London County Council Holborn-to-Strand improvement, and tramway-subway, *ibid*, 1910, **183**.

The main access tunnel at London Airport, *Concrete Constr. Engng*, 1952, **47** (Mar.).

Granter, E., Park Lane improvement scheme: design and construction, *Proc. Instn Civ. Engrs*, 1964, **29** (Oct.); *see also* discussion, 1966, **35** (Dec.).

Finn, E. V. *et al.*, Strand Underpass, *ibid*, 1966, **35** (Nov.); *see also* discussion, 1967, **37** (Aug.).

Road Tunnels Committee Secretariat:
  Documentation and Studies, 13th World Congress, Tokyo, 1967
  Documentation Digest, 14th World Congress, Prague, 1971
  Documentation Digest Additive, 15th World Congress, Mexico, 1975
  Permanent International Association of Road Congresses, Paris.

Peck, R. B., Deep excavation and tunneling in soft ground, *Proc. 7th Intnl. Conf. Soil Mech. Fndtn Engng*, Mexico, 1969, **State of the Art**.

Sverdrup & Parcel and Associates, *Cut-and-cover tunneling techniques*, 1, *A Study of the State of the Art*, 2, *Appendix*, U.S. Department of Transportation Reports FHWA-RD-73-40 and 41, (NTIS, PB 222997 and 8), 1973.

O'Reilly, M. P., Some examples of underground development in Europe, *Laboratory Report 592*, Transport and Road Research Laboratory, 1974.

Symposium, Cut and cover or tunnelling: the practical aspects, Concrete Society, May 1974.

Jobling, D. G. and Lyons, A. C., Extension of the Piccadilly Line from Hounslow West to Heathrow Central, *Proc. Instn Civ. Engrs*, 1976, **60** (May); *see also* discussion, 1976, **60** (Nov.).

Consulting Engineers Group, *Prefabricated structural members for cut-and-cover tunnels*, 1, *Design concepts*, 2, *Three Case Studies*, U.S. Department of Transportation Reports FHWA-RD-76-113 and 114, (NTIS, PB 273530 and 1), 1977.

Dasgupta, K. N. *et al.*, The Calcutta rapid transit system and the Park Street underground station, *Proc. Instn Civ. Engrs*, 1979, **66** (May); *see also* discussion, 1980, **68** (Feb.).

Coulson, C. R. and Stubbings, B. J., Alternative methods for cut and cover construction in congested areas, paper F2, *Mass Transportation in Asia Conference*, Hong Kong, May 1980.

McIntosh, D. F. *et al.*, Hong Kong mass transit railway modified initial system: design and construction of underground stations and cut-and-cover tunnels, *Proc. Instn Civ. Engrs*, 1980, **68** (Nov.).

2

Submerged Tunnels

2.1 DEFINITIONS AND APPLICATIONS

A submerged tunnel is essentially a subaqueous tunnel constructed by the prefabrication of units and their submergence and jointing in a prepared trench.

The descriptive term 'immersed tube' is also commonly used, but as 'immersion' has perhaps some suggestion of temporary submergence, and as a 'tube' is most typically of circular section, it seems preferable to adopt the wider terminology. The lengths in which it is prefabricated are here termed 'units', which are aggregated into the whole, rather than 'elements' which suggest diverse components of an integrated whole.

The construction techniques have very little in common with those of bored tunnelling, although the completed structure must meet the same fundamental requirements for traffic cross-section, gradients, alignment, ventilation, lighting and environmental amenity. The specialised skills employed are no longer those of tunnel mining—excavation and support—but of structural fabrication in steelwork or reinforced concrete, of handling floating units in a waterway with precise control of buoyancy and stress patterns, and of establishing uniformity of support and watertight joints.

Most submerged tunnels have been built for highways, some for railways and metros, and a few for sewers, water intakes and other uses. Many of the major tunnels are in estuarine waters under shipping channels and it is to such tunnels that the discussion principally refers, although the method is also widely used under rivers and canals and in harbours and offshore. Where the diameter is small, the techniques merge into those of pipe laying. It is normal for a submerged tunnel to be sunk below bed level, but it is possible for the structure to be partly or wholly above the bed, with a protective embankment covering it, provided it does not constitute a hazard to shipping and is not itself vulnerable.

2.1.1 Balance of advantage

Very sweeping claims are sometimes made for submerged tunnelling in comparison with bored tunnelling for subaqueous crossings, and there is no doubt that in many cases it is the right method. There are, nevertheless, important exceptions where submerged tunnels are impracticable and bored tunnels are appropriate. This subject is further discussed later in this chapter.

There are in particular two major aspects where this system will frequently offer substantial advantages, namely, in requiring shallower depth and in demand on less specialised resources.

2.2 TUNNEL DEPTH

Tunnel depth under a navigable waterway is determined by: (1) depth of channel and any intended dredging to greater depth; (2) depth of protective cover over the crown of the tunnel; (3) roof thickness; (4) headroom above the running surface or other design datum; and (5) construction depth below the running surface.

(1) The specified depth of channel for navigation is the same whether for bored or submerged tunnel. It is normally defined as the depth at low water and will apply over the whole channel width accessible to shipping.

(2) The submerged tube usually benefits in that the requirement for cover over the finished structure need not usually exceed about 1.5 m, which gives protection against ships' anchors. The nature of the existing bed is not very significant in this context, unless there is a particular risk of scour, because suitable material can be deposited to form the protective layer. In a bored tunnel, on the other hand, not only is it necessary to have at least similar protection against anchors and scour but there must also be a sufficient thickness of competent ground above excavation for safe working at the face. This depth is likely to be at least a few metres, and probably 10 m or more.

(3) The roof thickness of a circular tunnel is usually rather less than where a flat slab must span over two or three traffic lanes.

(4) For a bored subaqueous tunnel a circular section is almost invariable. A majority of American submerged tunnels are also circular tubes, but in most European practice a rectangular form has been adopted. In highway tunnels particularly, the circle implies a substantial extra height above the traffic clearance line, and even although the space can be profitably employed for ventilation, the total depth from crown down to roadway makes the road level that much lower.

(5) Similarly, the depth of invert below road level is greater in the circular form: but this does not affect road level.

These three factors — dredged depth, cover, and structure down to road level — therefore fix the level to which the tunnel must descend, and they normally show an advantage of several metres to the submerged tunnel. This in turn allows the approach gradients from the portals either to be made shorter with a reduction in the length of tunnel construction or to be made less steep with corresponding benefit to traffic using the tunnel. It may be remarked that for tunnelled metros there may be no such saving in tunnelled length and gradients.

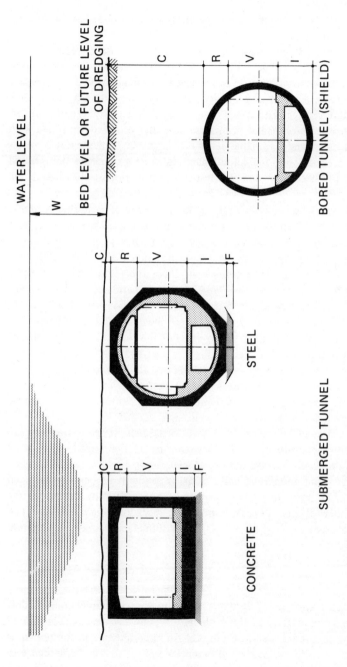

NOTE: 'W' = DEPTH OF WATER,   'C' = MIN. SAFE COVER,   'R' = ROOF CONSTRUCTION

'V' = VEHICLE CLEARANCE,   'I' = INVERT CONSTRUCTION,   'F' = THICKNESS OF PREPARED BED

Fig. 2.1 – The figure shows the comparative depths required for tunnels of similar function but differing forms of construction.

2.3 RESOURCES

The different demands on resources for the alternative methods have ultimately to be compared in terms of cost and time.

For the submerged tunnel the constructional work is mostly executed in the open, in shipyard or dry dock, with few inherent limitations on the rate of progress, and the different operations can proceed simultaneously. The necessary labour is drawn from a wider and less specialised field than in a bored tunnel. In contrast, virtually all construction of the bored tunnel is executed underground, where working space and access are severely restricted, and men with specialised skills not widely available must be employed. This restricted access limits the rate of progress possible. In many subaqueous tunnels compressed air working is found necessary, and this imposes further limitations on available manpower and possible rate of progress.

A submerged tunnel scheme may require the provision of a suitable drydock, and also the design and manufacture of specialised floating craft or gantries with control equipment for placing and jointing the units. For a single short tunnel the time and cost of special equipment might be unduly burdensome, but for longer tunnels and for successive tunnels in the same region most of these installations can be re-used. Even very large items of floating construction plant can be towed a long distance to a new site, and so can units built in a drydock, or in a shipyard. For the Chesapeake Bay crossing the cylindrical steel units were fabricated in Texas and towed 2700 km to site. For the Hong Kong metro tunnel the special construction platform was built in Japan and towed to Hong Kong and back again, a voyage of similar length, taking two to three weeks.

Another aspect of making the best use of resources is in the choice of the number of traffic lanes to be provided initially. It is common for the traffic demand to be predicted as for two lanes only at first, but with the expectation that there will ultimately be a need for additional lanes. If the extra lanes are built immediately there will be no financial return on the extra cost incurred until traffic has increased beyond the two-lane capacity. For this reason a policy of building a two-lane tunnel initially, with subsequent duplication, is often adopted. If four or more lanes will ultimately be needed the capital outlay will be least if the whole project is built in a single operation. Against this must be set the interest charges incurred on the extra cost of the facilities not immediately essential.

2.4 HISTORY

The idea of prefabrication and submergence as a method of tunnel construction was discussed in London early in the 19th century, and a scheme was promoted for a crossing of the Thames at Rotherhithe. An experiment was conducted at Horseferry Stairs as early as 1811 when two brick units of 2.7 m diameter 7.6 m long were built and placed in the river bed. Owing to difficulties of interference with navigation in the limited period of slack water at high tide and problems of finance, the scheme proceeded no further, and Brunél's shield-driven Thames tunnel was eventually built, followed later by the development of the Greathead shield. The combination of shield and compressed air has been found effective for all subsequent tunnelling under the river Thames, but submerged tunnels are currently being considered.

Ideas for submerged tunnels were also put forward and discussed at various times during the 19th century in America as well as in England and France. The first substantial construction of this type appears to have been for a sewer in Boston Harbour. In 1893 sections of sewer, 2.7 m outside diameter and in lengths up to 20 m, were built on shore, using brick lining in a steel skin, and were floated out and sunk in a prepared trench, to form an inverted siphon across a channel, the units being connected with external bolted flanges. In the following year for an outfall the section was modified to comprise 200 mm brickwork, 150 mm concrete and a 100 mm timber skin.

In 1906-10 a tunnel for the Michigan Central Railroad was constructed to replace a ferry crossing of the Detroit river which connects Lakes St. Clair and Erie. The approaches were tunnelled with shield and compressed air, but for the river section of 812 m the softness of the clay and depth of 17 to 24 m made this method appear hazardous. A system of construction in trench was accordingly devised. Twin cylindrical steel shells, 7 m diameter x 80 m long coupled transversely with steel diaphragms and furnished, outside the proposed external wall concrete, with timber formwork, were prefabricated in a shipyard, closed at the ends with temporary bulkheads and floated to site. At site they were sunk, with the aid of pontoons and water ballast, onto preset foundation grillages in a prepared trench. End joints were made as steel lap joints carefully fitted at the shipyard and bolted up by divers. Concrete filling in the trench bottom and in the external walls and roof of the tunnel was deposited by tremie. Internal concrete lining was a subsequent operation.

It will be noted that the prefabrication was minimal, of the watertight membrane and external shuttering only, and that the concreting was all done *in situ*. It was perhaps more closely related to cut-and-cover, but under water, than to modern submerged tunnelling, but it did embody the germ of the method.

This set the pattern for cylindrical steel units and shipyard fabrication which was followed in the Detroit—Windsor tunnel in 1928-30 and subsequently at most other American east coast crossings. Circular tunnels were also adopted on the west coast, but as concrete structures built in drydock, for the Posey Tube (1925-28) and Webster Street Tube (1957-62) at San Francisco. The Bay Area Rapid Transit tunnelled crossing of the Bay was a concrete shell with two circular tubes and ancillary ducts within a steel hull. The typical American highway crossings are constructed with a single cylindrical tube of about 11 m diameter carrying two traffic lanes.

Rectangular submerged tunnel construction was initiated in Rotterdam for the Maas tunnel (1937–41). This was a highway tunnel incorporating dual two-lane carriageways and also cycle and pedestrian subways. The foundation problem of providing satisfactory bearing over a wide flat area was overcome by developing a system of sand injection beneath the units. The units, 61 m long, were partially built in drydock, floated to a jetty for completion, and sunk onto temporary supports. Steel plates were incorporated for waterproofing. Steel piling and concrete and compressed air working were used when closing the gaps. Jointing by the use of rubber gaskets between units was developed first for Deas Island tunnel, Vancouver (1959), and is now standard practice, largely dispensing with the need for compressed air or divers' work.

It is no accident that this particular form of tunnel has developed so vigorously in the Netherlands, stimulated by the geographical pecularities, although it was Danish engineers who initiated the special techniques. Three rivers, the Rhine, Maas and Scheldt, have created deltas through which they discharge to the North Sea by numerous channels, supplemented by canals, and with great cities and ports growing astride the rivers.

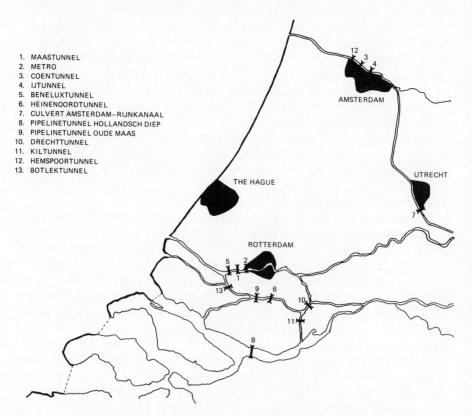

1. MAASTUNNEL
2. METRO
3. COENTUNNEL
4. IJTUNNEL
5. BENELUXTUNNEL
6. HEINENOORDTUNNEL
7. CULVERT AMSTERDAM–RIJNKANAAL
8. PIPELINETUNNEL HOLLANDSCH DIEP
9. PIPELINETUNNEL OUDE MAAS
10. DRECHTTUNNEL
11. KILTUNNEL
12. HEMSPOORTUNNEL
13. BOTLEKTUNNEL

Fig. 2.2 – Principal submerged tunnels in the Netherlands.

Bridge construction has been handicapped by the need for wide and high spans for shipping and by the difficulties of foundations in soft alluvial ground. Deep bored tunnel crossings have likewise been considered unusually difficult because of the softness of the ground, but the relatively shallow submerged tunnel of rectangular form has shown great advantages, and the country's available resources in shipping and docks and navigational skills have been put to good use. The growth of these subaqueous tunnels in Holland may be compared with the sequence of Alpine tunnels in Switzerland nearly a century earlier.

The circular and rectangular sections have continued to be typical of American and European practice respectively, but each has benefited from devices adopted from the other, and the choice, in any particular case, is strongly influenced by available resources, and experience of the particular system.

2.5 SEQUENCE OF OPERATIONS

The principal operations in construction are:

1. Initial fabrication
2. Trench preparation
3. Fitting out, floating in and sinking
4. Jointing, bedding in and backfilling

The content and relative importance of the various operations differs according to whether the initial shell is of steel or reinforced concrete and whether it is of circular or rectangular form. There is no more a standard routine for this method of construction than for bored tunnels; each site presents its own problems and opportunities.

The differences in procedure between the use of steel cylinder and concrete rectangle are so marked that it is clearest to deal with the details of each method separately. Nevertheless, the general principles are clearly the same, and there is much common ground. The particular techniques of jointing and of bedding, although typical of one system or the other in practice, are not essentially so, and are now adopted as considered most suitable for the individual site. Steel cylinders and their traditional techniques are discussed first, followed by rectangular concrete tunnels.

2.6 STEEL CYLINDERS

2.6.1 Initial fabrication
The basis of design is shipyard fabrication of a cylindrical steel shell, but the finished tunnel usually comprises a reinforced concrete structural lining and a road deck, within a cylinder about 10 m in diameter of 8 mm thick steel,

surrounded by concrete contained within a thin octagonal steel casing. The outer concrete, for which the steel octagon acts as formwork, provides external protection against damage and corrosion, and necessary ballast, contributing also to structural rigidity. (See Fig. 2.3.)

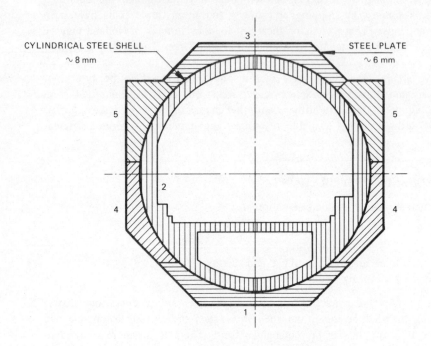

SEQUENCE OF CONCRETING

1. Structural concrete applied as a keel before launching
2. Structural reinforced concrete
3. Structural concrete added before sinking
4. Ballast concrete for sinking operation
5. Ballast concrete added by tremie after sinking

Fig. 2.3 – Typical cross section of American tunnel.

In the shipyard the steel cylinder is welded up and stiffened by transverse diaphragms connecting it with the external octagon. Watertightness of the welding is most important for the finished tunnel. A unit about 90 m in length is assembled and a concrete 'keel' is cast between the inner and outer steel casings. The open ends are closed off by temporary steel or timber bulkheads.

At this and all subsequent stages buoyancy of the units has to be carefully controlled. The weight of the steel cylinder, octagon and diaphragms may be of the order of 5 tonnes per metre and the concrete keel and bulkheads can bring this up to a launching weight of 15–20 t/m, but the water displacement weight of the cylinder is about 80 t/m. Initially, therefore, it will float one quarter submerged.

Before launching, reinforcing rods, for subsequent use, are placed within the cylinder and access hatches are provided. Sideways launching is preferred, but endways is practicable if stresses are properly allowed for.

After launching further external concrete is added including a cap to contribute to longitudinal stiffness, and concrete in the sides as necessary ballast for towing requirements.

Each end of the unit is furnished with devices for subsequent underwater connection between units as further described below.

Stresses have to be analysed for each step in construction. Longitudinal bending stresses are induced by uneven loading when floating. Initially, the end bulkheads must be considered as point loads on a beam, and subsequently the sequence of concrete placing must be planned to avoid excessive stresses. The thin steel cylinder must not be subjected to circumferential stresses which could cause buckling. Stresses must also be examined where loads may be imposed in towing and at points of support during sinking.

Fabrication of steel cylinders can be advantageous in effecting early completion of a project, because manufacture and delivery of units can be started quickly and arranged to comply with any reasonable sinking programme. If site work is delayed storage afloat of units is not likely to be difficult.

2.6.2 Trench preparation

The trench in the river bed must be excavated by dredging or by grabbing; if rock is encountered blasting may be required. Special dredging equipment is likely to be necessary not only because dredgers currently in use at a port normally operate only to the depth of the shipping channel, whereas an excess depth of 15 m or so will be required, but also because harbour dredgers are unlikely to be idle and available for any but brief periods.

It is usual to excavate the trench to about 0.6 m below the underside of the units to be sunk and with sides sloped back at a stable angle which may be about 1 : 3. The bottom has then to be filled up with gravel, screeded accurately to level. This screeding operation presents considerable difficulties at depth in a waterway. Heavy steel screeds must be controlled to follow the specified gradient and must be independent of tidal movements which affect all floating craft. Deposition of silt in the trench may make it necessary to clean out each length immediately before placing the gravel and again before landing the unit.

The cylindrical unit is relatively stiff longitudinally, and the narrow base of the external octagon can be bedded satisfactorily on the screeded surface without imposing unacceptable stresses on the structure.

Where ground conditions are variable and there is a risk of uneven settlement, piled foundations can be employed. In the Hampton Roads tunnel in Chesapeake Bay average settlement of 0.22 m has been measured but without unacceptable differential movement. It is attributed principally to rebound of soft silts, after trench excavation and before placing, followed by recompression.

2.6.3 Fitting out and sinking

The operation of fitting out will include completion of the inner concrete lining and road deck and of the external concrete, together with attachment of sinking equipment, such as masts, pontoons, winches and jacks. Some of this may be done at the shipyard site if it is reasonably near the construction site, but a special jetty may be necessary. The unit must continue afloat until required for sinking. The necessary balance of weight and buoyancy for the sinking stage may be maintained by the use of pontoons, the weight of the unit being slightly in excess of flotation, or the unit may remain self buoyant, requiring the addition of ballast, but finally it must be given a factor of safety against uplift.

At the fitting out site, the concreting of lining and roadway through hatchways will be completed, together with enough of the outer concrete shell to ensure minimum freeboard of 0.15 m or so. This brings the weight to about 99% of the displacement weight, which, for a 90 m unit, may be about 7000 tons with the final weight when in place exceeding this by about 600 tons. As much work as possible is done at the surface before sinking, and therefore pontoons or other means of temporary support are normally employed to keep the unit afloat while sinking weight is increased to about 300 tons net. Floating cranes, winches or block and tackle can be used for controlled lowering.

Position is carefully regulated by six anchor ropes, one fixed across the line from each corner and one each ahead and astern. Target masts are usually fitted on top of the unit so that level and position can be kept under continuous survey from fixed stations.

The net weight of the submerging unit is only about 300 tons and this load can be handled easily but the total weight is over 20 times as much and its inertial mass and that of the surrounding water may give rise to heavy dynamic loads in the supporting and anchoring ropes, particularly if any rapid movement develops and has to be checked. Very steady conditions are therefore essential during sinking.

Another factor which may be of great importance in tidal waters is the salinity of the water and any changes in it. Sea water has a specific gravity of about 1.025, which increases the buoyancy of a 7000 ton unit by 175 tons. Obviously the average salinity and any changes during handling and sinking of a unit must be studied.

2.6.4 Jointing and backfilling

The ends of each element are fitted with cylindrical steel sleeves projecting through square end plates and accurately made to match the adjacent units. Cover plates outside the sleeves project further, round the lower half of the circle at one end and the upper half at the other, so as to form a lapped joint when finally connected. Locating brackets project sideways at the ends of the horizontal diameter to be connected together with large diameter bolts. There may be further tensioning devices to pull the adjoining ends together (Fig. 2.4).

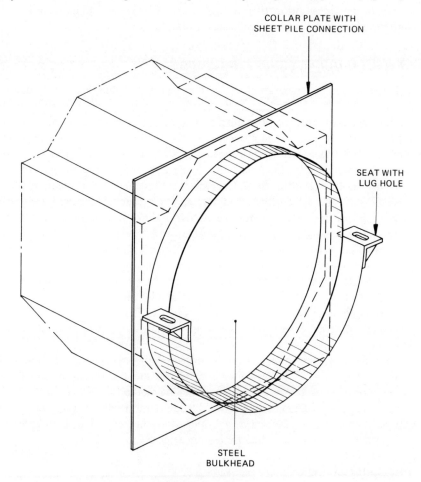

COLLAR PLATE WITH
SHEET PILE CONNECTION

SEAT WITH
LUG HOLE

STEEL
BULKHEAD

Fig. 2.4 – Perspective view of joint connection.

Divers are employed to make the connection, fit the bolts and caulk the gaps between cover plates and cylindrical sleeves. The unit is then further secured by careful placing of sand backfill under and against the sloping sides. The elaborate devices for the rectangular concrete units described below are not necessary. The vertical edges of the square end plates are fitted with sheet pile interlocks to take a vertical semi-cylindrical closer plate, within which concrete is placed by tremie to encase the whole joint. This completes the external sealing of the joint sufficiently for the bulkheads to be cut out at the appropriate time and for the steel tube to be finished off by watertight welding, followed by lining concrete.

The backfilling of the trench and protective cover over the top complete the operation. In addition to guarding against damage by shipping, the cover contributes added stabilising load against flotation.

2.7 RECTANGULAR CONCRETE TUNNELS

A similar sequence of operations is followed for rectangular reinforced concrete structures, but the detailed procedures differ significantly. The total weight of the finished structure per unit length will be approximately proportional to traffic capacity because it is governed by the requirements of buoyancy. The fabricated steel in a cylindrical steel unit, however, constitutes less than 10% of the final weight and is in itself structurally stiff and watertight, the concrete providing additional stiffness, and necessary ballast and also protection to the steel shell. In the reinforced concrete units, whether rectangular or circular, most of the concrete must be completed before the unit functions as a structure which can be handled. In practice, the rectangular tunnels have most often been built with four carriageways as compared with two for the steel tubes and also in lengths sometimes 50% greater. As compared with weights of 500 tons for the steel tube alone and 7000 tons for the finished unit 90 m long, a rectangular four-lane unit 140 m long might weigh about 28,000 tons.

Watertightness is a second aspect in which different techniques have been developed, although originally the concept of the welded steel plate membrane was adopted.

A third major difference is in the provision of uniform support under the wide flat base; the sand jetting equipment devised for the Maas tunnel was considered to be the key to this whole tunnelling method.

The concrete rectangle does not readily lend itself to steel lap joints between units, and compressed rubber joints have been developed progressively in a manner that does not ordinarily require attention by divers.

2.7.1 Initial fabrication

As already mentioned, a casting basin or dry dock is required for building of the units to the stage of floating out.

If an existing dry dock can be adapted the units can be constructed one or two at a time, but the operations of flooding and then pumping out again are time consuming and costly. It is more usual to excavate a special casting basin in which all of the units, or possibly half, can be built before any are floated. A very large flat casting area is necessary at a depth exceeding 8 m below high water. A sound foundation must be assured for each unit, preferably surfaced with gravel and possibly with other devices to ensure that water pressure has free access under the base at the floating out stage. Concreting plant has also to be installed on a very large scale.

In an existing dock, gates will be used for access to the main waterway, but if a basin is specially constructed for a single use, some more temporary device, possibly with sheet piling, will be required, and an access channel of adequate depth and width. Controlled flooding arrangements must be provided. In typical estuarine conditions ground water lowering throughout the construction period will probably be necessary to keep the dock dry and prevent heave of the bottom.

It is now usual to complete the construction of the unit except for some ballast concrete, in the dry dock. Buoyancy is of critical importance at all stages as has already been discussed for steel tubes. A weight of 99% of the total displacement weight with a freeboard of about 0.1 m is appropriate for the floating and fitting out stage, and finally after sinking an excess weight of up to 10% to give a factor of safety against uplift. These requirements determine closely the proportions of the tunnel cross section in terms of: (a) clear space; (b) wall, roof and floor thickness; and (c) ballast concrete. The specific weight of the finished concrete as actually constructed must be very accurately determined both for the high grade reinforced concrete in the structure, and the ballast concrete.

For two-lane and four-lane units at moderate depth these conditions can be met with a reinforced concrete cross section but, for a three-lane span and at depths exceeding perhaps 15 m or under high embankments, transverse prestressing may be necessary, because the structural depth in the unit cannot be increased without losing buoyancy. On the other hand, prestress in the unit before its working load is imposed may make necessary added reinforcement or other counterbalance.

Width of units is of course determined by traffic capacity required, and height primarily by traffic clearance plus concrete thickness. Additional space, if necessary, for ventilation or other purposes, may be provided laterally, but a small requirement for extra space to accommodate booster fans may sometimes be met by increased height. The length of units will be the maximum practicable for handling in the dock and waterway. In the original Maas tunnel the length was 61 m but in current practice 100 to 140 m is favoured and units up to 268 m have been used. Long units effect economy in jointing, but manoeuvrability in handling in the waterways must be maintained. Long units also present problems of stresses in handling and of shrinkage and cracking. They may lack flexibility in accommodating slight settlement on their foundations.

As already noted, overall waterproofing has developed from the original welded steel membrane. Designers have adopted combinations such as a steel plate for the base and a bituminous membrane in walls and roof, or wholly bituminous membranes. Systems have now been devised dispensing entirely with membranes and relying on the concrete alone, except that construction joints in the casting process are sealed with waterstops cast in.

In current practice the whole base slab is usually cast first in lengths with upstanding waterstops to seal the necessary construction joint between base and wall. The wall is cast in sections of 15 to 25 m to minimise shrinkage cracks, the construction joint again being crossed by a water stop. Where membranes are dispensed with, each concrete section is cooled artificially during the setting process of the concrete to extract the heat of hydration and minimise shrinkage cracking caused by the temperature differential between mature base and fresh wall concrete. Unless a very high standard of workmanship and control can be ensured, the bituminous or other membrane is still likely to be preferred.

The roof section may be cast at the same time as the walls, or may be a separate subsequent operation, the construction joints again being fitted with waterstops, for which special epoxy resin grouting techniques have been adopted.

The vertical construction joints between sections are designed, if necessary, to accommodate slight movement where settlement may be expected. Rigid units are necessary for handling in the towing and sinking operations and in the final adjustment of position. If a flexible design has been adopted it is made temporarily rigid until finally in place, either by longitudinal post tensioning cables or by reinforced concrete links across the construction joints in which the steel ties can subsequently be cut.

2.7.2 Trench preparation

There is no great difference from the procedures for steel cylinders previously discussed. The trench will normally be wider but a little shallower. An excavated depth of 0.6 m below the unit is considered appropriate. This space is not filled until after the unit is in place, when sand is jetted in. If the river carries much silt it may accumulate in the trench and will have to be cleaned out immediately before depositing the sand fill. The same jetting system may be adapted, first to blow out the silt and then to place the sand.

2.7.3 Fitting out and sinking

When the units, or batch of units, have been completed in the dry dock, they are fitted with end bulkheads, and, as appropriate, alignment towers and temporary access shafts, before filling of the dock. There are usually water ballast tanks provided in each unit to allow its buoyancy to be controlled by pumping. In many cases units are allowed to remain wholly or partly submerged in the building dock until required, only one at a time being floated and finally equipped for sinking. Pontoons are the most usual equipment for regulation of sinking, but floating cranes may be used or special jack-up platforms. Where the units are kept a little short of buoyant, pontoons may be floated into the dock over them, to pick them up with winches, the pontoons and units being transported as a single entity to the site of the trench, but this is dependent on ample depth of water in dock, exit channel and main waterway.

Sinking procedures are much the same as for steel cylinders; although the gross weight may be three or four times as much, the net weight is kept quite low, but dynamic and hydrodynamic loads must be very carefully monitored and kept under control. Motion of the unit cannot be changed rapidly.

In any but the simplest conditions of towing and placing, tests on models in a hydraulic laboratory are advisable to allow estimates of the necessary towing and anchoring power to be made and to predict the behaviour of the units in currents. Of particular importance are tidal currents possibly with varying salinity and therefore density. The final position of the unit will almost inevitably be across the flow of the current, and in a narrow channel, the unit will form a significant obstruction giving rise to substantial hydro-dynamic forces, and possibly causing scour in the channel.

At least six tugs are likely to be required to bring the unit into position for sinking. Lines from winches on the unit must then be fixed successively to the anchorages set in the river bed to allow the unit to be brought precisely into position above the trench. Lowering may be effected by first submerging the unit with the aid of water ballast, then lowering it from the pontoons very nearly to its final level and position, and after a final check landing it on its prepared points of support.

It is usual to land the unit on a preset beam or on foundation blocks at its outer end and preferably on a bracket on the previous unit at its inner end. Final adjustments both vertical and horizontal may be made by jacks working from the beam, but in recent practice the horizontal adjustment is often made by jacks acting longitudinally on the side walls at the joint under construction.

The net ground pressure from these tunnels is so low that a sand fill foundation is usually considered to give rise to little risk of significant settlement, although in soft silts there may be appreciable general subsidence. In a few exceptional cases piles have been adopted as the final foundation, and accurate level and bearing have been established by injection of grout between the heads of the piles and the bearing beam.

2.7.4 Jointing and backfilling

In connecting two units the initial water seal is established by use of a rubber joint gasket which allows the internal steel and concrete work at the joint to be completed. The rubber seal, having the so-called 'gina profile', is bolted to the face of the unit being placed and projects about 160 mm in all. The outer tip is in section a 40 mm triangle of soft rubber backed by the main body of the gasket about 160 x 120 mm of harder rubber. When the unit is positioned the nose of the gasket bears on a steel plate on the fixed unit and is pulled in

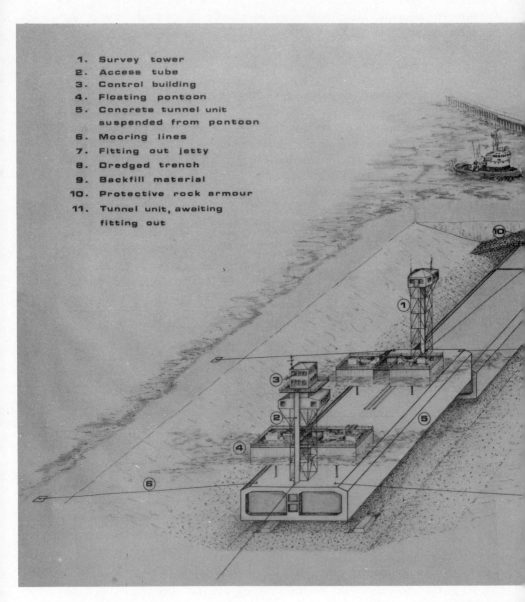

1. Survey tower
2. Access tube
3. Control building
4. Floating pontoon
5. Concrete tunnel unit
    suspended from pontoon
6. Mooring lines
7. Fitting out jetty
8. Dredged trench
9. Backfill material
10. Protective rock armour
11. Tunnel unit, awaiting
    fitting out

Fig. 2.5 – Submerg

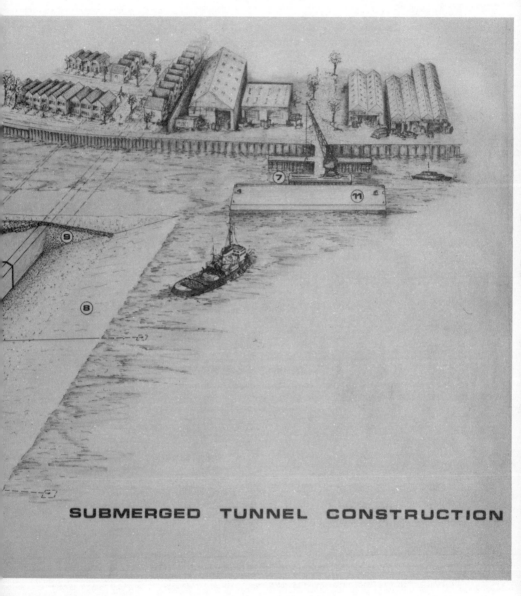

SUBMERGED TUNNEL CONSTRUCTION

unnel Construction.

towards it mechanically so as to compress the soft rubber tip and form a water seal.

A closed compartment is thus formed between the bulkheads of the adjoining units. The enclosed water which is at a pressure corresponding to the depth, is drawn off or pumped out, releasing the pressure and leaving unbalanced the pressure on the bulkhead at the far end. This forces the units into closer contact at the joint and compresses the main block of hard rubber. If necessary, to prevent excessive movement in compression, steel limit stops are fitted.

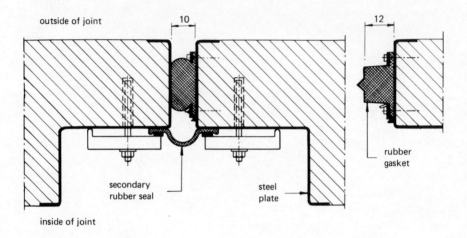

Fig. 2.6 — Joint seals for concrete units.

A steel 'omega' plate can then be welded across the gap inside the protection of the rubber gasket, and the bulkheads can be removed. The concrete lining is completed between the two units, the detail depending on the extent and nature of any future deflections which are to be accommodated.

2.7.4.1 Sand filling

The development of a technique to provide a solid well compacted sand fill beneath the tunnel units was the key to the whole system of rectangular construction. A turbulent mixture of sand and water is prepared on a barge, or elsewhere, and pumped down to nozzles beneath the unit. The jetting system can be adapted to remove accumulated silt before placing sand.

In its original form a complex special gantry was used to carry the jetting nozzles and to move and direct them so as to traverse systematically the whole area to be covered. At the same time as jetting sand in, excess water was sucked out through a suction nozzle, allowing the sand to settle as a densely compacted mass. One disadvantage of this system is that the gantry, located over each unit in turn, constitutes a continuing obstruction to shipping.

It has now been established that in favourable conditions fixed vertical nozzles, underneath the base of the unit and at a spacing of about 4 m, can inject sand satisfactorily. At the initial stage of injection the sand settles out as a conical deposit, spreading in all directions, until a central crater is formed where turbulence maintains the sand in suspension. A succession of narrow escape channels through the rim are eroded and, after contributing to the spread of the cone, become choked again. It is found in practice that virtually the whole base area can thus be given necessary support. The vertical jetting pipes may pass through the floor of the tunnel, being fed from within or, preferably, the supply pipe system may be carried round the side of the structure, being supplied from a mixing barge. It appears that the grading of the sand is not at all critical provided it is not so coarse grained and porous that water leaks away from the crater and turbulent flow is not maintained.

When support of the base is assured the sides of the trench are backfilled and the top of the tunnel is covered to the prescribed depth.

2.8 TOPOGRAPHY

The topography most favourable to the choice of a submerged tunnel is where the river crossing is wide and shallow with a sandy bottom, and with gently sloping banks and channel, and currents are not great. Depths have not hitherto exceeded about 30 m. The choice between bridge and tunnel for a new crossing has first to be the subject of careful study, which probably involves outline design, narrowed down progressively to details of type and technique. A precise balance can rarely be struck even in terms of cost: there are uncertainties as to actual circumstances encountered, and benefits and environmental value cannot be quantified with any precision.

Assuming that a tunnel is proposed, the relative merits of bored and submerged tunnels require to be assessed for the particular case. The general advantages of a submerged tunnel have already been outlined but some of the special difficulties need to be further examined. Each can usually be overcome, but if several factors are unfavourable the cost may be excessive.

2.8.1 Geology
The geology of the river bed is of primary importance. If it is of soft alluvial strata to a sufficient depth below the required clear channel level, the necessary trench can probably be excavated satisfactorily, but if a thin layer of silt overlies a rock bed the problem becomes more difficult. At the least, heavy dredging

Fig. 2.7 – Submerged tube in Australia. Single extremely long concrete unit 23 m × 5.5 m × 263 m carrying cooling water to Eraring power station under Dora creek. Unit ready to float out preparatory to sinking to form the crossing.

(Electricity Commission of New South Wales)

plant capable of working below normal depths will be employed, possibly with the aid of blasting. In such ground, a bored tunnel may not need so great a depth of protective cover as in soft ground, and free air or low pressure working may be feasible.

2.8.2 River profile

The transverse profile of the river bed in section along the line of the tunnel may be very significant. If the slope exceeds the acceptable tunnel gradient, then a trench which is of minimum depth at mid river, or at the edge of the shipping channel, must be made progressively deeper as the bed level rises more rapidly than the tunnel. It may be noted that if the tunnel crosses the slope obliquely the effect is to provide a gentler slope than at right angles.

Where the waterway is bounded by a quay wall the dredged depth may extend to the face of the wall, where a vertical profile is presented. Excavation of the trench up to the wall may imperil its stability and excavation behind the wall will necessarily be very deep.

A high flood protection bank will similarly add locally to the depth of trench and therefore to its width unless the sides are supported. At the site of the Dartford tunnel these conditions would have required a trench over 30 m deep through a flood bank recently raised, and having only a small margin of stability. This was one among several factors favouring a bored tunnel.

The Elbe highway tunnel at Hamburg (1971) presented a similar problem where the north bank rises steeply to a height of 35 m. The crossing is for six lanes of traffic. A submerged tunnel with reinforced concrete units was constructed under the waterway but boring was adopted for the long rising gradient under the north bank, the transition being effected at a ventilation station.

2.9 TIDAL WATERS

Tides in an estuary may make the handling and control of floating units very difficult and impose a strict timetable. Where the tidal range exceeds 5 m or so the shipping channels may not be usable at low water except for vessels of very shallow draught. Most shipping will therefore move up and down river near the time of high water which will also be the time for access through locks into and out of docks. This will also be the only period when adequate depth is available for towing tunnel units to site.

During sinking of each unit restrictions on shipping will be essential with complete closure of the channel for at least one tide and probably for 24 hours. In a tidal channel where there is a danger of a unit grounding while being towed to site, the operation should be effected on a rising tide, and to a strict and carefully devised timetable. Contingency planning is essential against the risk of grounding and thereby blocking the channel. Safe deepwater anchoring points may be identified, to be utilised in the event of unforeseen delays.

Once the unit is in place over the trench the operations of anchoring, sinking and final location may take about eight hours. Sand filling and subsequent removal of equipment may take several days, during which some restriction of shipping may continue to be necessary even if the channel is very wide.

Tidal currents in river estuaries may be very strong, and are likely to vary in direction and ultimately reverse as the tide ebbs and flows. Flood water in a river system may further complicate navigation. Powerful tugs adequate in capacity and numbers are essential for handling the units, which will be particularly difficult to control when turned across the channel ready for anchoring and sinking.

Salinity and therefore buoyancy may vary during the rise and fall of a tide, and it sometimes occurs that a front of advancing salt water pushes forward as a wedge beneath river water flowing downstream at the surface. A further phenomenon in tidal rivers may be the slow transport seawards of a relatively thick layer of silt at the bottom. This would be trapped by the tunnel trench and must be removed before backfilling under the tunnel.

In some estuaries the combination of large tidal range, convergent shore lines and a steeply rising bed, may result in the formation of a 'bore' like that of the river Severn in which the rising tide advances as a very steep wave front, which would obviously be a serious hazard in the handling of tunnel units.

2.10 ADVANTAGES

Provided that the difficulties can be overcome, the submerged tunnel of rectangular form gains added advantage when more than a two-lane tunnel is required. The cost of dry dock construction, if it is needed, is spread over the larger project and the trench is excavated with the same plant and less than double the volume of excavation. The longer the submerged tunnel, of course, the less significant is the heavy cost of a special casting dock and other special plant and equipment.

The advantage of shallow depth and short approaches up to ground level can be of overriding importance in city development where land is closely built up and in some cases space for access roads and ramps is more readily available in the immediate vicinity of the river.

Where the choice has been made in favour of steel cylinders rather than concrete it is obvious that a fundamental factor is the availability of low cost fabricated steel. The extra depth required may be quite acceptable, and speed of construction is likely to be much greater, partly because of the time required for preparing and equipping a suitable dock. Cylinders normally accommodate two traffic lanes only and therefore the rectangular form gains some advantage when four or more lanes are required as they can be embodied in a single unit with one joint, where the cylinders require two. Steel cylinders can more easily be made seaworthy for towing in exposed waters.

2.11 CHANNEL TUNNEL

The Channel Tunnel is in itself a subject of interest, but is of particular relevance in this chapter in exemplifying the limitations of existing methods and possibilities of development.

From early in the 19th century various forms of submerged structure have been suggested for a Channel tunnel, some laid on the sea bed and others in trench. It would certainly be essential that the structure should be in trench at a safe depth below possible damage by ships, and that any section above bed level should have very substantial protection. In the context of a rail tunnel the subject is discussed in Volume 1, but if a highway tunnel is to be contemplated, different considerations arise. Ventilation necessary for road traffic is not practicable for a full length tunnel without shafts, but much steeper gradients are acceptable for road than for rail, and bridge-tunnel schemes are suggested with shorter tunnelled lengths under the shipping lanes connecting with bridged lengths at artificial islands. In such a scheme the width required for multi-lane carriageways and ventilation ducts is a factor favourable to submerged unit construction. One scheme proposed embodies roadways on viaduct from each coast to offshore islands, with a central section in submerged tunnel ventilated by shafts up to artificial islands, all in combination with a through rail tunnel for the whole length.

The whole project, however, is on a scale so much greater than any existing submerged tunnel that every detail would have to be analysed critically. It would be very dangerous to accept as practicable any aspect because it is merely an increase of scale in a proved technique.

A factor favourable to the method is the great length of tunnel which offers the advantages of repetitive fabrication and handling, increasing experience and maximum use of equipment. In contrast, the great length to the working face of a bored tunnel is a severe logistical handicap. Access through the length of the submerged units would also be necessary for jointing and adjustment and the necessary ventilation might be difficult.

Among the principal adverse factors are the depth, the problem of gradients, the open sea exposure and the intensive international shipping.

Depth greatly exceeds that of any existing submerged tunnel for traffic, and the relationship between structural depth and maximum stress in slabs and the requirements of buoyancy might well preclude the use of a rectangular concrete structure, necessitating a circular shell, or arched form. On the other hand, temporary buoyancy might be provided by attached pontoons or special floating craft.

Depth also adds to the difficulties of making the structure watertight, particularly at the joints between units. Any access by divers for jointing or remedial work would be at pressures exceeding 5 atmospheres, requiring special modern diving techniques. Depth would also make more difficult precise control of placing by the long ropes needed for lowering and for anchors, so special plant would have to be devised and developed to function reliably.

For a railway tunnel, imposing strict limitations on gradients, the contours of the sea bed may present a problem. In normal surface construction in undulating country the required profile is based on a balance of cut-and-fill, working to the preferred gradients. For the submerged tunnel, embankments, except perhaps very locally and on a small scale, are undesirable as being difficult to build, vulnerable to shipping, and possibly subject to erosion and subsidence. Where the slope of any undulation exceeds the limiting gradient, increased depth of cutting becomes necessary with disproportionate increase of excavated volume. The surface technique of resorting to bored tunnelling where cutting depth becomes excessive is not available.

The disposal of material excavated from the trench and the procurement and placing of fill would present considerable problems of handling methods and cost.

The exposed conditions in the open sea are not altogether without precedent as in the Los Angeles sewer outfall and in offshore oil operations, but they are unprecedented in the handling and placing of units such as these. In the English Channel the tidal range is of the order of 5 m and tidal currents are strong and changeable. Stormy weather is frequent, with unpredictable wave action. The problems of precise location become formidable, even more so in fog. No doubt radar controls, laser beams and other devices could be adapted and developed, but at mid-channel, over 15 km from land, they would need to be very accurate and reliable, including projection downwards through 50 m of water.

Shipping in the Channel is international and includes almost the whole range of seagoing craft from Channel swimmers up to very large oil tankers. Regulation and control of such shipping in sufficient detail to safeguard anchored construction craft, in new positions as each unit was placed, would demand powerful and effective authority, not presently established. The hazards to shipping from any bridge-tunnel scheme would necessitate elaborate precautions, and also effective control.

It is not surprising that the estimates of cost and the uncertainties of feasibility have hitherto ruled out this technique for the Channel tunnel, but undersea oil exploration and exploitation have led the way in developing new methods and equipment, and it may well be that such a scheme will become practicable and competitive.

BIBLIOGRAPHY

For up-to-date details of submerged tunnel construction, and for historical records, it is necessary to refer to papers published in the proceedings of engineering institutions and of specialist conferences and also to articles in the technical press.

The earliest examples in the United States and Holland respectively are described in:

Kinnear, W. S., The Detroit River Tunnel, *Trans Amer. Soc. Civ. Engrs*, 1911, **74**.

Wilgus, W. J., The Detroit River Tunnel, *Min. Proc. Instn Civ. Engrs*, 1911, **185**.

Van Bruggen, J. P., The Road Tunnel under the River Maas at Rotterdam, *Engng.*, 1940, **150** (Aug. 9 and 30, Sept. 27).

Important general papers on the subject, which include historical material and more extensive bibliographies are:

Brink, A., Recent developments in the design of submerged tunnels, *Struct. Engr*, 1966, **44** (Feb.); *see also* discussion, 1966, **44** (Aug.).

Bickel, J. O., Trench type subaqueous tunnels: design and construction, *Struct. Engr*, 1966, **44** (Oct.); *see also* discussion, 1967, **45** (Feb.).

Pequignot, C. A., Selective bibliography on immersed tubes, *Tunnels and Tunnelling*, 1969, **1** (July).

Brakel, J., Some considerations of submerged tunnelling, *Proc. Instn Civ. Engrs* 1971, **48** (Apr.); *see also* discussion, 1972, **51** (Jan.) (esp. Prangnell, K. J., pp. 142–144, on J. I. Hawkins' experimental tunnel under the River Thames at Rotherhithe in 1811.)

Williams, G. M. S. and Innes, K. W., Structural aspects of submerged tube tunnel construction, *Struct. Engr*, 1972, **50** (Feb.); *see also* discussion, 1972, **50** (Aug.).

Palmer, W. F. and Roberts, K. C., Developments in trench-type tunnel construction, *J. Constr. Divn., Proc. Amer. Soc. Civ. Engrs*, 1975, **101** (Mar.).

Glerum, A. *et al.*, Motorway tunnels built by the immersed tube method, *Rijkswaterstaat Communications*, 1976 (25).

Brakel, J., *Submerged Tunnelling*, Technische Hogeschool, Delft, 1978.

Immersed Tunnels, *Delta Tunnelling Symposium*, Amsterdam, 1978.

Many of the tunnels have been described individually in some detail. References to them will be found in the above general papers and in the general bibliography in this volume.

3

Shafts and Caissons

3.1 WORKING SHAFTS

For many tunnels sinking of a shaft to provide working access is the first operation. Even where access for progressive excavation from a portal or from an adit is possible, shafts are usually advantageous in any but short tunnels in providing additional points of attack and more favourable working sites, as well as avoiding initially the complexities of portal construction.

A working shaft may be sunk as such and subsequently abandoned, but most often the shaft will have a permanent function and its siting and design will be such as to serve both purposes.

Working shafts are typically vertical but may be inclined, in which case special safety precautions against runaway vehicles are necessary. Neither in terms of construction nor of function is there a clear demarcation between an inclined shaft and an inclined tunnel. For escalators, at a slope of 30°, the term 'shaft' is widely used, perhaps because it replaces the old lift shaft.

The fundamentals of construction of shafts are very similar to those of horizontal tunnels, and may be compared briefly. Excavation is safer in that there is no overhead ground to be taken out, but all spoil has to be lifted and hoisted, without delay, to clear the working surface. Immediate support of the exposed ground generally presents less difficulty than in a tunnel, while for permanent support the same systems are available and linings are more easily placed. Management of water is of major importance: all water entering finds its way quickly to the shaft floor unless intercepted, and unexpected water presents an immediate threat of flooding. Another water hazard is heave of the bottom due to upward water pressure from below an impervious stratum. It may be noted that shafts have to be sunk through whatever ground happens to overlie the tunnel they serve; there is usually some choice of stratum for the tunnel, but not for the shaft.

A working shaft is ordinarily used for all construction access, but special shafts may sometimes be provided for such temporary functions as conveyor disposal of spoil, or for ventilation. Where compressed air working from the earliest stages is intended the shaft must accommodate an air lock. Shafts for tunnel construction do not often extend to the great depths of hundreds, or even thousands, of metres required for mining. The greatest depths are probably in hydroelectric work, including pumped storage schemes, to convey water at high pressure to the turbines. In long rail or road tunnels in mountain areas, intermediate shafts for ventilation may have to be deep but depths may be excessive, as in the Mont Blanc highway tunnel where ground cover exceeds 1000 m over 80% of its length of 11.6 km, and there are not intermediate shafts.

In many of the early English railway tunnels speed of construction was sought by sinking shafts at short intervals so that many tunnel faces could be worked simultaneously. From each shaft two faces could be worked, by miners and bricklayers alternately. In the very early Kilsby tunnel (1838) no fewer than 18 shafts were sunk to drive a tunnel 2.2 km long. Wider spacing, up to 1 km, was more usual subsequently. Of course, smoke clearance was another reason for provision of intermediate shafts.

For metro systems, which are constructed as near as practicable to the surface, shafts are correspondingly shallow. Permanent shafts are usually necessary in the vicinity of stations and can be sunk at an early stage with adaptations as appropriate for the tunnelling operations. Suitable sites for intermediate working shafts are often difficult to obtain in built-up areas.

Subaqueous tunnels are usually driven from a shaft or pit on each side of the waterway. Brunel, for his Thames tunnel, sank a single 15 m diameter brick shaft to a depth of about 17 m and launched the shield from the bottom. For the first tube of the Mersey Kingsway tunnel (1971) shafts were sunk on each side of the river at the sites of the ventilation stations and were used for four pilot tunnel drives, riverwards and landwards, but the main tunnel was driven from the south portal, cutting straight through to the site of the north ventilation station.

3.2 REQUIREMENTS

The principal requirements for an all-purpose working shaft are:

1.   Availability from an early stage of construction until completion. The sinking of a shaft, from the bottom of which tunnelling can be started, is often the first site operation. If the shaft is ultimately intended for permanent use, its use for construction access may have to be given up at some stage as work proceeds, but adequate alternative access can then usually be secured.

2.   The minimum acceptable area depends on the length and diameter of the tunnel to be driven but, except for a short length of small diameter tunnel, a shaft diameter of about 4 or 5 m is the least that will accommodate reasonably hoisting space for materials of construction and equipment, and a safe ladder bay for man access. A diameter of 6 m or more is much to be preferred,

especially where the depth and extent of the work calls for a separate man hoist. For a shield driven tunnel the shaft must provide space for lowering the shield components, and possibly for their assembly, although a separate shield chamber as an enlargement of the tunnel is often preferred, and is of course unavoidable where the shaft does not lie on the centre line of the tunnel.

The shaft must in any case be of sufficient diameter for the construction of a breakout at the bottom, whether for the actual tunnel or an access passage. Its diameter should preferably not be less than 1½ times the diameter of the breakout, because it is very difficult to support adequately the opening formed by the intersections of two nearly equal cylinders.

3. The working site must be reasonably level and have space for a crane, whether a fixed derrick or a travelling crane, and for storage of materials and of spoil. The materials to be stored include a stock of timber and steel for temporary support, tunnel segments or other permanent lining materials, cement and aggregate. Heavy ground loads, as from stacks of segments, should not be close to the perimeter of the shaft. For spoil disposal from a restricted working site a minimum stock pile is necessary to maintain the balance between output and disposal. An overhead hopper on a gantry is often used in city sites.

4. There must be good road or rail access to the shaft for delivery of materials and disposal of spoil.

5. The first working shaft has to be used to carry down the required tunnel alignment from the surface survey network.

In very many cases, and almost invariably in urban and subaqueous tunnelling, alluvial deposits, often water-bearing, overlie the chosen tunnelling stratum which may be clay or rock. The upper part, at least, of the shaft must therefore be excavated and supported through soft ground. In urban conditions it is likely to be of particular importance to minimise ground movements and strains which could result in settlement and cracking of adjacent buildings. The construction techniques must be decided with this in mind. In urban conditions, also, noise and other inconveniences during shaft sinking, tunnelling and in permanent use, must be kept within acceptable limits.

It should be appreciated that such relatively shallow shafts, required only during the construction period, are not comparable in magnitude with deep mining shafts, where heavy capital costs are necessarily incurred. Savings in time, permitting earlier mine development, and in winding costs, for large volumes of material over a long period of use, can justify substantial expenditure on special sinking techniques and equipment. The permanent headgear may be installed and utilised for the sinking operation. Diameters, commonly 6 m to 8 m, allow for ventilating air for the mining work as well as hoists for men and materials. The choice of methods and expedients for tunnel work must be adapted to dates of site access, availability of equipment and suitability for the ultimate permanent use of the shaft.

3.3 PERMANENT SHAFTS

In addition to provision of temporary construction access, shafts may be permanently required as elements in the tunnel system for many purposes, such as:

1. Lifts: (a) originally common for passenger access to metros, but now largely superseded by escalators; (b) access to underground power stations.
2. Escalators: inclined shafts, usually at 30°, for passenger access to metros and also to pedestrian and cycle subways.
3. Stairs and ladders: principally for emergency access and for maintenance.
4. Ventilation: (a) natural ventilation of main line and metro rail tunnels, and draught relief; (b) forced ventilation of highway tunnels, metros and rail tunnels.
5. Conveyance of liquid: (a) as part of the main flow in hydroelectric pressure tunnels; (b) collecting surface water for conveyance by tunnel; (c) drop shafts in sewers, where steep gradients are not acceptable; (d) undersea sewer outfalls; (e) undersea intake or outfall for cooling water for power stations.
6. Pipes and cables in river crossings.
7. Drainage and pumping, particularly from subaqueous tunnels.

   Shafts are also constructed for purposes not directly ancillary to tunnelling:

8. Mining shafts (which have already been referred to).
9. Temporary storage and treatment of sewage.
10. Bridge and other deep foundations.
11. Hydraulic lift pits.
12. Wells.

3.4 WATER IN SHAFTS

Many shafts, perhaps a majority, are sunk through water-bearing ground and the management of this water is a most important factor in sinking technique. In urban tunnelling alluvial deposits must usually be penetrated to reach the chosen tunnelling stratum which may be clay or rock. In subaqueous tunnels soft water-bearing strata are likewise to be expected, and the shafts may be started through open water in a cofferdam or caisson. The principal methods of handling such water problems within the shaft are: (1) pumping, from a sump in the shaft kept ahead of the main excavation; (2) ground water lowering, either by well points or deep well pumping; (3) grouting of pores and fissures; (4) freezing; and (5) compressed air working.

The principles of these methods are discussed in Vol. 2, Chapter 5, Ground Treatment, and Vol. 1, Chapter 5, Compressed Air Working, but particular applications are further referred to below. High water pressures at depth impose a limitation on pumping, groundwater lowering and compressed air working.

3.5 SINKING TECHNIQUES

Techniques of sinking differ principally in the manner of providing support for the ground. More than one method may be utilised for a single shaft. In sound dry rock immediate ground support may be unnecessary, particularly if smooth walls can be ensured, as by reaming. The principal methods of shaft construction and lining include: (1) timbered pit; (2) sheet piling and secant piling; (3) concrete walling in slurry trenches; (4) precast segmental linings; (5) *in situ* concrete lining; (6) caisson sinking, in free air or compressed air; and (7) rotary drilling.

Whatever method be adopted for the deeper part of the shaft, the initial work, on land, is nearly always the levelling and preparation of the working site, the laying down of a kerb defining and protecting the perimeter of the shaft and the provision of a crane. Where the shaft has to be sunk through open water the first step will be construction of a staging, if practicable, and a sheet piled cofferdam or, alternatively, the use of a caisson built from a staging *in situ* or floated out.

3.5.1 Timbered pit

The principles of timbering in support of the ground in sinking a shaft vertically are very similar to the support in a horizontal heading. In soft ground excavation may be taken to considerable depths even when wet by the use of piling or forepoling. Poling boards are driven down ahead of excavation and secured by heavy walings as space is excavated. A pumping sump is maintained ahead of general excavation. Excavation thus proceeds in stages governed by the length of the poling boards in each setting.

Where the ground is suitable the sides of the initial excavation from the surface are often supported by timber runners or steel sheeting set vertically behind a timber or steel frame and driven down as excavation proceeds. This allows a depth of several metres to be taken out as the initial operation and simplifies the timbering compared with forepoling. Such a pit may serve as a starting point for a deeper permanent shaft of circular form with cast-iron or concrete lining. Apart from the advantages of an early start to work with readily available resources, complex permanent works may ultimately be required at the head of a shaft and the temporary timbered pit is readily adapted to requirements for later completion.

3.5.2 Sheet piling and secant piling

A shaft of limited depth may be wholly accommodated within a steel sheet piled

Fig. 3.1 – *In situ* concrete lining forming collar at top of shaft. (*E. Nuttall Ltd.*)

excavation whether on land or in water. The cofferdam can be made of any desired size and shape to accommodate headworks for the shaft. Depths exceeding 20 m or so make it progressively more difficult to ensure straight driving and secure interlocking of the piles. If the shaft has to descend into rock there is liable to be leakage of water at the soil/rock interface, and a sufficient margin of space between shaft and cofferdam must be left for appropriate water sealing devices in addition to allowing for possible deviation of the piles.

Fig. 3.2 – Sheet piled shaft with timber framing.

Steel sheet piling is obviously very useful where soft water-bearing ground of limited depth overlies a stratum into which the piles can be driven, for example in initiating a shaft to be driven through water-bearing gravel above a bed of clay in which the main tunnel is sited. It is also a most important method for intake and outfall shafts to be sunk under water.

Secant or contiguous bored concrete piles may serve the same function as steel sheet piles in preforming the shaft walls in soft ground.

In metro work the whole area of a station is sometimes excavated from the surface in a single pit, the running tunnels being then driven out from or into the bottom.

3.5.3 Slurry trench walls

This is an alternative to sheet piling or a timbered pit in relatively shallow shafts or shaft heads through soft ground, particularly granular soils. Narrow trenches are excavated, where shaft walls are required, by accurately controlled grabs, the sides being supported by the use of bentonite suspensions or other mud slurries. No timbering is required to retain the sides of the trench. Concrete to form the shaft wall is deposited by tremie pipe in the slurry-filled trench and displaces the slurry, which can be recovered for re-use. Shaft walls can thus be constructed, with reinforcing steel if necessary, before the main excavation is started, and the time taken and obstruction caused by timber strutting is avoided. Straight walls are usually implicit in the trenching techniques, and an octagon may be constructed to accommodate a finished circle.

The method is perhaps most useful in urban conditions. There is less danger of settlement in adjacent ground and structures and the noise and vibration of pile driving are avoided. Sound concrete requires careful supervision, and if a good surface finish is required a secondary lining will probably be necessary.

3.5.4 Precast segmental lining

Lining with circular cast-iron tubbing is one of the historic methods of shaft sinking and can be used to great depths, with increased thickness of lining where necessary for higher pressures. Precast concrete, pressed steel or welded steel segments are used similarly. The relative merits are discussed in Vol. 1, Chapter 8 on Permanent Linings.

A great advantage of segmental lining over *in situ* concrete is that in wet and difficult ground each segment can be set in place with a minimum of excavation and the whole ring will immediately develop its full strength in resisting compression as soon as all its segments are in place.

The description following is of a C.I. lined shaft in soft ground, sunk by underpinning from a prepared level which may be at the surface or at the bottom of a pit or cofferdam. It is assumed that excavation is by hand or by grab. Open excavation to a depth of a few rings is frequently possible with the bottom accurately levelled off and prepared for the building up of C.I. rings to above starting level. A supporting collar embodying if necessary a steel joist or other platform is constructed to secure the rings against any settlement as excavation progresses. The rings may be built up to above ground level to form a safety barrier. Shaft sinking proper then starts by the technique of underpinning. If the ground continues dry and firm, excavation for each ring may be taken out completely, possibly with a little timbering locally at the sides, followed by assembly and bolting up of a ring. It may be noted that, unlike tunnel rings, a special set of 'key and two top' segments is usually unnecessary. The shaft segments are identical and closure can be effected by taking out a little extra excavation behind the last pair of segments, which allows them to be fitted in from the outside. It is then important to seal the bottom of the

Fig. 3.3 – Ventilation shaft with concrete walls cast in slurry filled trenches.

Fig. 3.4 – 8 m shaft with concrete segmental lining being sunk through clay by underpinning.

ring and grout up the back of the ring, not only to seal against water but to fill any voids and prevent settlement of the ground and uneven pressures on the ring. It is essential at all times to ensure that the weight of the completed lining is adequately supported and cannot slide down.

Where the ground is loose or water-bearing, particularly in granular soils, it becomes undesirable to excavate the whole area of the shaft for a depth of one ring before erecting any lining. By local excavation of a small pit each segment successively can be set in place and bolted up, strutted back if necessary from the 'dumpling'—the ground left undisturbed in the centre of the shaft. A drainage sump can be kept ahead of the general excavation with the water level controlled by pumping or baling. Early and thorough grouting is even more important where water is present.

In a cast-iron or concrete lined shaft compressed air working can be adopted if necessary by provision of an air deck at the head of the shaft and an air lock for entry. With compressed air the water level can be pushed down to the bottom of the completed lining. It is of great importance in such a shaft to ensure that the air deck is designed to resist the maximum working pressure, and is sealed against leakage which may be through honeycombed concrete, or joints affected by shrinkage. In the shaft itself all joints will require caulking and bolts will require grummets.

Cast-iron is lacking in tensile strength and the shaft should not be subjected to tension from the uplift pressure of the compressed air. Kentledge should be added above the deck accordingly. Circumferential tension should also be minimised by grouting.

The general principles of compressed air working are described in Volume 1, Chapter 5.

In deeper shafts water pressures may be too high for such measures to be adequate. Pumping from external deep wells or from boreholes sunk through the floor of the shaft may effect adequate lowering of ground water levels. Advance grouting, or freezing, may be found necessary.

3.5.5 In situ concrete lining

In dry rock which is, at least temporarily, self-supporting, *in situ* concrete has substantial advantages in avoiding the problems of manufacture, delivery and storage of segments, and in being readily adaptable to varied conditions of overbreak. It has the disadvantages of not providing immediate strength, and of requiring formwork, which has to be erected, filled, left in place for the time required by the concrete and then stripped out and lowered for the succeeding lift. *In situ* concrete is probably the most widely employed lining for deep shafts. Excavation and lining have to be executed from a set of stagings, lowered as the work advances, equipped for drilling, for loading and hoisting and for shuttering and concreting. Excavation is commonly by drilling and blasting, the spoil being loaded into hoisting skips by a grab or mechanical

Fig. 3.5 — Deep rock shaft, prior to *in situ* lining. Working platform suspended. (*E. Nuttall Ltd.*)

shovel. The operation is cyclic and not continuous, and it is of great economic importance to ensure a properly balanced cycle of operations minimising loss of time. Where rock conditions permit, lifts of about 8 m for the lining are common, against which there may be four or more cycles of excavation. Ventilation adequate to clear the blasting fumes quickly is important.

Water poses the usual difficult problems and may be controlled by grouting or freezing. It is, of course, necessary to protect freshly placed concrete against flows of water, which might wash out the fines. Separation of lining from rock and provision of drainage channels is sometimes effected by insertion of corrugated sheeting against the rock. There is also the problem of thermal insulation where concrete is to be placed against frozen rock, both to avoid frost damage to the concrete while setting and to avoid thawing of the surrounding ground by the heat of the concrete.

3.5.6 Caissons

A caisson is essentially a watertight box, of which the base at least is prefabricated, set in position and sunk there by excavation from within. The walls and any required structure may be built up progressively as the caisson is sunk. The caisson may be open and be sunk by open grabbing, or it may be decked across at any desired level to provide a working chamber capable of retaining compressed air and equipped with an air lock. An open cylinder such as that used by Brunel to form the working shaft for the Thames tunnel is the same in principle. So is a 'monolith', the term being generally descriptive of a substantial concrete box, as used for quay walls, sunk into place by open grabbing.

A caisson may be used to form a shaft either on land or through water. It may also be constructed to embody a section of tunnel of special complexity, such as the ventilation supply and exhaust shafts and their connections into the tunnel, as in the Dartford and second Blackwall tunnels. In the Amsterdam metro a difficult length of tunnel in the city was constructed by sinking a series of caissons subsequently connected with the aid of nitrogen freezing; one reason for use of the method was to minimise settlement of the adjacent ground.

An open caisson is built up from a cutting edge set in position at the surface or in a pit. The walls are built up vertically together with any necessary cross walls. In dry ground men can excavate at the bottom out to the cutting edge, and the caisson sinks under its own weight and that of any added kentledge. Control of sinking may be aided by jacks or winches. In wet conditions excavation may be entirely by grab until a firm bottom is reached and penetrated by the cutting edge, which allows the caisson to be pumped dry. Sinking in these conditions is more difficult to control, and the caisson may finish out of position and out of level. Some allowance for this is necessary in the original design.

The use of compressed air can greatly assist precision and control and minimise loss of ground causing settlement in the vicinity.

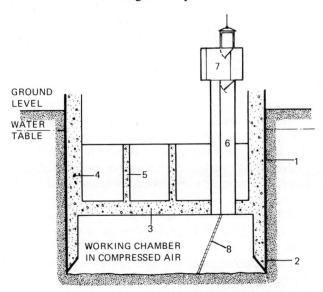

GROUND
LEVEL

WATER
TABLE

WORKING CHAMBER
IN COMPRESSED AIR

1.	Steel shell	5.	Internal structure as required
2.	Steel cutting edge	6.	Steel 'figure 8' access shaft with —
3.	Ait tight deck	7.	Ait locks for men and materials
4.	Walls built progressively during sinking	8.	Access ladder

Fig. 3.6 – Compressed air caisson.

The caisson must be provided with an airtight deck, usually 2.5 m or so above cutting edge level, through which passes a working shaft capped by an air lock. This forms a working chamber within which men can excavate, the spoil being hoisted in skips up the shaft and through a muck lock. Initially, excavation in free air may be practicable but, as the head of water externally increases, compressed air working becomes necessary. Water pressure can be balanced precisely at cutting edge level, unlike the conditions at the vertical face of a tunnel. The structure within the caisson will be built up as sinking proceeds, but careful regulation of weight is always necessary. There must be sufficient weight to overcome skin friction on the sides of the caisson, but there must be no risk of it descending out of control. In tidal waters the necessary balancing of water pressure by air means that at high water there is added uplift. Kentledge is often necessary at some stages of the work, or water ballast, which can be pumped in and out, may serve. Lowering of the air pressure, after a shift has completed excavation and come out, is known as 'blowing down' and may be a regular procedure. It can greatly assist sinking but needs to be done under control if the caisson is not to lurch suddenly and move out of position. For deep narrow shafts in particular, frequent careful surveying is needed to know what is the exact position and attitude of the cutting edge and to allow instructions to be given for excavation to be regulated accordingly.

Resistance to sinking, particularly skin friction, is not easily predictable, and may be unexpectedly high. Caisson walls are sometimes built to a slight inward taper, or stepped in above the cutting edge with the intention of reducing friction. Bentonite slurry injected round the perimeter has been found to be a useful lubricant.

The final setting to level can be done with reasonable precision by first constructing within the working chamber suitable concrete blocks finished to the proper level, with any appropriate allowance for bedding in. The final blowing down allows the caisson to settle on to those blocks, but it will probably be necessary to restore the pressure again for the concrete filling of the remainder of the working chamber, followed by grouting up.

It has been assumed in the above description that a separate working chamber beneath the level of any other structural work is provided, but it is, of course, possible to site the air deck in the caisson at a higher level, and to construct whatever is required within this much deeper working chamber. This may allow the caisson to be founded at a somewhat lesser depth but it makes control during sinking much more difficult with an unnecessarily large volume of air and a more top-heavy caisson.

Fig. 3.7(a) – Working chamber in caisson at Dartford Tunnel. Exploratory drilling in progress.

Fig. 3.7(b) – Caisson at Dartford Tunnel. Structure above working chamber
forming shield chamber for reception of tunnel shield.

Fig. 3.7(c) — Excavation at cutting edge of caisson.

3.5.7 Drilling

Various drilling methods for shafts have been developed in recent years, including mud slurry methods and offshore techniques used in drilling for oil. Large diameter augers have been used in clay, at relatively shallow depths, but probably more for foundation work and for site investigation than for shaft construction. In mining work shaft raising in rock by boring and reaming machines has become established but diameters have usually been limited. Experience with TBMs can usefully be applied to such shaft work.

The provision, installation and removal of any such boring machines is a costly operation and unless the required shaft is a deep one or there are special circumstances, it is unlikely that the expenditure can be justified. Except for undersea outfalls or intakes, few tunnel access shafts have therefore utilised these methods. An exception is the St. Gotthard highway tunnel, where two of the four deep ventilation shafts were constructed by machine boring. The shafts in question were inclined at about 40° to the horizontal and were bored upwards at a diameter of 3 m and then enlarged downwards to the required 6.6 m, the lengths being 860 m and 500 m through paragneiss and granite.

In drilling shafts downwards, a major problem is the clearing of the broken rock from the face being drilled. Provided that satisfactory fragmentation of the rock can be obtained, circulating mud slurry can be used to remove the spoil. The mud slurry can also serve the purpose of supporting the sides of the excavation until excavation is complete, when a steel cylindrical lining can be lowered in and sealed and grouted into place.

The merit of raise boring is that the fragmented rock falls away from the face, but the method depends on having proper access at the base of the proposed shaft for drilling and removal of spoil. In raise boring it is common first to drill a central pilot hole and then to ream this out to the full required diameter. The pilot may be drilled either up or down and likewise the enlargement. If the enlargement is drilled downwards, spoil can be discharged through the pilot. It is obvious that if uncontrolled water, or unconsolidated strata are encountered in drilling up, difficulties may be very severe.

In shaft sinking to a depth of about 200 m, a method employed has been mud drilling of the required shaft followed by lowering with the aid of water ballast of a cylindrical steel liner sealed with a base plate. The cylinder was extended in 5 m lifts as it sank and was finally grouted in.

3.6 SHAFT BOTTOMS

There are various special considerations in constructing the bottom of a shaft, such as precautions against heave, sealing of the base, provision of a sump and pumps, and provision for breaking out for tunnel construction.

Even in a relatively shallow shaft water pressure at the bottom level has to be considered during the final operations and after completion. The base slab must be designed to resist uplift by its own weight and by anchorage into the lining of the shaft. Although the shaft may finish in a stratum which is impervious this may well be underlain by one which carries water under pressure, even a very thin seam of sand. If there is any reason to suspect this, a test boring through the bottom of the shaft is advisable. It may be feasible to relieve the pressure during construction of the base plug by allowing the water to rise through the test bore, with a gravel and sand filter if necessary, and to handle the water with the ordinary sinking pumps. A more elaborate deep well pumping procedure might be needed, or grout injection. Where uplift pressure is only marginally in excess of loading, the shaft bottom could be concreted in a series of trenches so that ground weight is not lost except over a narrow strip until replaced by concrete.

The base slab, already mentioned, has to be capable of resisting uplift and should normally be made watertight by grouting. It may also have to accommodate various items of construction plant and permanent equipment. In water-bearing ground provision of a sump and pumps is almost invariably essential both for the construction stage and in service.

The appropriate sump capacity may well be such that the whole shaft is sunk to below the level of the adjoining tunnel, and this will allow subdrains to be provided below the tunnel invert to assist in construction. In river crossings where the tunnel dips to its deepest level at mid river, drainage shafts are sometimes sunk to below this level and connected to the lowest point by special drainage headings rising gently from the shaft.

3.7 TUNNEL EYE

Another important feature of shaft construction is provision of a tunnel eye from which to break out. Where the shaft lining is segmental the cross joints in the rings at tunnel level should be arranged to suit the intended opening and minimise any cutting of segments or unnecessary exposure of ground. The operation of breaking out requires great care, especially because the excavation has to be done in ground already disturbed by the sinking of the shaft, and because water from higher levels may well follow down the outside of the shaft. A practice adopted in soft ground is first to make a small opening in the shaft and from this to drive a short timbered heading. At a distance of a few metres a 'break-up' for the full size tunnel is excavated. In this break-up chamber two or more rings are built and grouted and the tunnel is then built one ring at a time back to the shaft. Alternatively, two rings may be built within the shaft followed by advance one ring at a time through the shaft wall. The shaft rings when built carry the external ground pressure and in breaking them out the loads must be redistributed, initially to temporary steel or timber framing, and ultimately to the new tunnel lining.

Where a single segment is omitted, or removed, redistribution of load into adjacent rings may be expected, but as the opening is enlarged timber struts, wedged into each ring before removal of a segment, can take the load temporarily. The final junction between tunnel and shaft is likely to comprise steel joist framing embedded in concrete.

3.8 SUBAQUEOUS SHAFTS

There is an increasing demand for intake or discharge of water in rivers, lakes or the sea for power station cooling water, hydroelectric intakes, or sewer outfalls. In water of moderate depth the use of a cofferdam allows normal methods to be adopted, but special problems in constructing shafts and connecting them into tunnels arise in deeper water or in exposed situations or on a hard rocky bottom. A combination of expedients is likely to be necessary, dependent in some degree on whether or not the shaft is to be provided with headworks and to be capable of subsequent closure for maintenance.

For example, at the Wylfa Power Station cooling water intake the coast was exposed and the sea bed at -11 m was rocky with a tidal range of about 6 m. The

Fig. 3.8 – Tunnel eye. Framing at bottom of circular shaft to accommodate access tunnel.

shaft headworks were constructed from a jetty by sinking a pit, by blasting in the sea bed and concreting in a cylindrical shell and bulkhead. The lower part of the shaft was completed by raising a 1.2 m diameter shaft from the end of the tunnel and then enlarging to 4 m diameter downwards. In contrast, the Dublin outfall sewer at Howth Head, where strong currents were expected, was constructed almost entirely from the end of the tunnel below. In sound rock the shaft was raised from the end of the tunnel nearly to the sea bed and, after flooding the top of the shaft, was blasted with a carefully prepared pattern of charges, a sump having been prepared to absorb the debris. In fact, subsequent construction of headworks by divers proved necessary because of progressive silting up with sand. A similar method was employed for a cooling water outfall in the Bay of Fundy.

Other such shafts have been driven by employing offshore drilling rigs and boring down and capping off at or a little above bed level, then driving the tunnel into the shaft bottom and removing the cap.

BIBLIOGRAPHY

Much of the general material published relates to very deep shafts for mining development. The less deep shafts associated with tunnelling are mostly described as incidental to specific projects. They include shafts for tunnel construction, access, ventilation, hydroelectric power, marine outfall and intake. Caisson sinking is also used extensively for bridge and other deep foundations. A few relevant references are listed here, with further papers in the general bibliography.

Wilson, W. S. and Sully, F. W., Compressed-air caisson foundations, *Instn Civ. Engrs Works Construction Paper 13*, 1949.

Symposium on Shaft Sinking and Tunnelling, London, 1959.

Chapman, E. J. K. *et al.*, Cooling water intakes at Wylfa Power Station, *Proc. Instn Civ. Engrs*, 1969, **42** (Feb.); *see also* discussion, 1969, **44** (Nov.).

Haswell, C. K., Thames Cable Tunnel, *ibid*, 1969, **44** (Dec.); *see also* discussion, 1970, **47** (Oct.).

Collins, S. P. and Deacon, W. G., Shaft sinking by ground freezing: Ely Ouse – Essex scheme, *ibid*, 1972, **Suppl.** p. 129; *see also* discussion, p. 319.

Various authors, Shafts, *Proceedings Rapid Excavation and Tunneling Conferences*:
  2nd, San Francisco, 1974, Session 16, Chapters 68–72;
  3rd, Las Vegas, 1976, Session 3, Chapters 8–11;
  4th, Atlanta, 1979, Sessions 14 and 15, Chapters 68–76.

Lander, J. H. *et al.*, Foyers pumped storage project: planning and design.

Land, D. D. and Hitchings, D. C., ditto: construction, *Proc. Instn Civ. Engrs*, 1978, **64** (Feb.); *see also* discussion, 1978, **64** (Nov.).

4

Geology

4.1 RELEVANCE

Of the natural sciences geology is that with which tunnelling is most intimately involved — from the inside of the rock mass. Tunnel engineering is concerned with those aspects of physical geology which are descriptive of the ground surrounding the tunnel, its response to stresses imposed and existing stress patterns. Geotechnology is a branch of civil engineering dealing particularly with this field, and relating the structural and mechanical behaviour of soils and rocks as affected by engineering construction to the geology in its widest sense. Its principal subdivisions are: soil mechanics, rock mechanics, hydrology and seismology.

There is inevitably an overlap between the subjects discussed in this chapter and elsewhere, more particularly in Chapters 3, 4, 5 and 6 of Volume 1.

It is mainly in geological terms that the ground can most accurately and concisely be described as an integrated and intelligible whole, both in surveys and reports, and in records of construction. Standard geological maps provide much valuable information, but inevitably need to be supplemented by site studies, to locate precisely and evaluate the significance of important features such as discontinuities, and by quantitative assessment of mechanical properties of the rock.

A brief discussion of the structures and vocabulary is therefore appropriate here, with a description also of the subject matter of soil and rock mechanics, not attempting to expound their principles.

4.2 SURVEY REQUIREMENTS

The tunneller wishes to know what strata he may expect to encounter, and how ground water is likely to manifest itself. He requires an assessment of the probable behaviour of the ground as and when it is excavated, including the effects of disturbance by the processes of excavation and drainage. In considering disturbance, the time factor is very important both in the initial response of the ground during construction and the subsequent relaxations and stress changes after the permanent lining has been installed. The requirement is to have defined in as much detail as practicable the nature of the strata and their geometry in all directions round the tunnel. The detail survey should cover a width at least several times the tunnel diameter and a depth from the surface to well below the lowest invert level foreseen. The description of the ground in general geological terms needs to be supplemented by analysis of its small scale structure in the terms of soil mechanics and rock mechanics, and by identification of the probable and possible locations of boundaries and discontinuities as precisely as may be.

The construction problems of excavation, support and water management, discussed in Vol. 1, Chapters 4 and 6, are all closely dependent on the composition and granular structure of the soils and rocks encountered and on the pattern of division by bedding planes, joints and faults. The broad division into soft ground and rock tunnelling is a useful one; at the extremes they are very different but an infinite variety of tunnel conditions and construction methods lie between. The geologist uses the term 'rock' for all natural deposits, although he does classify as 'unconsolidated sediments' much of the tunneller's 'soft ground'. In this chapter 'rock' is sometimes used in the comprehensive sense, but where discrimination is relevant 'rock' is applied to consolidated strata in contrast to 'soils' for unconsolidated sediments.

Much urban tunnelling is in soft ground, because cities are frequently situated on rivers with wide areas of alluvial deposits, and most tunnels, for urban transport and for sewers in particular, are constructed at as shallow a depth as practicable. In contrast, mountain tunnelling is likely to be through hard and varied rock which has been subjected to the stresses of mountain building earth movements. In geologically recent mountain systems, such as the Alps, and most other major mountain ranges, some strata may not be fully consolidated and may be under heavy continuing stress so that difficult and dangerous 'squeezing rock' is encountered.

There are two essential aspects of a geotechnical survey and report. These are:

1. Collection of all available information from site and other sources, and preparation of maps and sections to forecast what rocks and structures and ground water may be encountered.
2. Description and classification of the ground through which the tunnel will pass in terms of soil and rock mechanics so as to predict its behaviour in response to tunnelling operations.

It is of the greatest importance that the information should be presented in a form which can be utilised by the tunnel engineer in planning the project, if subsequent major changes required by unforeseen conditions are to be avoided. An adequate, clear and accurate statement of the available information and of

Position of Tunnel Face   Set No. 9

1. Degree of weathuring in the Silurian

	MUDSTONE				SANDSTONE (Grade strengthened weathering)		
Complete	Zone 1	✓	1)	Very weak	Complete	✓	
High	Zone 2	✓	2)	Weak	High	✓	
Moderate	Zone 3	✓	3)	Mod. strong	Moderate	✓	
Slight	Zone 4		4)	Strong	Slight		
Fresh	Zone 5		5)	Very strong	Fresh		

2. Ratio of Sandstone to Mudstone   1:10
3. Thickness of Sandstone Beds   50–300 mm
4. Details of Discontinuities

(continues overleaf)

		Joints		Bedding Planes	
Spacing					
Very Close	< 5 cm	✓		✓	
Close	5–30 cm				
Moderately Close	30–100 cm				
Wide	1–3 m				
Very Wide	> 3 m				
Orientation		Strike	Dip	Strike	Dip
Tightness					
Tight		✓		✓	
Medium					
Open		✓		✓	
Irregularity					
a) Degree of Planeness					
– plane		✓		✓	
– curved				✓	
– irregular					
b) Degree of Smoothness					
– slickensided					
– smooth		✓		✓	
– rough		✓		✓	
Filling Material					
Type					
Thickness					
Degree of Cementing					
Not cemented					
Weakly cemented		✓		✓	
Strongly cemented		✓		✓	

Fig. 4.1(a) – Example of geological log for rock tunnel face.

(continued from previous page)

5. Groundwater

6. Description of Dykes, Faults, Folds, Synclines etc
    Description of Foliation

7. Comments

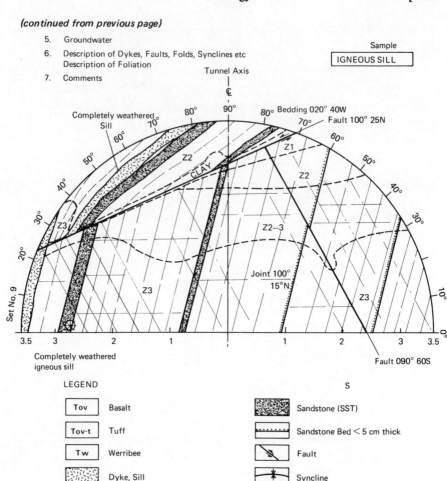

Fig. 4.1(b) – Example of geological log for rock tunnel face.

the uncertainties is essential to the production of an efficient scheme, including the proper choice of methods and plant to be used.

The geological information may be summarised as:

1. Lithology—type and characteristic of soil and rock.
2. Stratification and jointing.
3. Discontinuities—including faults and fissures.
4. Water.
5. Existing stresses.
6. Mechanical properties: (a) soil mechanics; (b) rock mechanics.
7. Seismic activity.

The manner in which some of these items appear in a tunnel face is exemplified by Fig. 4.1 which shows a tunnel log recorded for the Melbourne Loop. The form records lithology, weathering and discontinuities in sedimentary strata: steeply dipping weathered mudstone and sandstone. It exhibits also an intrusive sill and a normal fault. Frequent records of such detail at the face allow accurate plotting of the geological structure, which can be of the utmost value both currently and for future reference.

4.3 LITHOLOGY

Lithology describes the physical, chemical and textural characters of rocks and soils. These vary over so vast a range that classification and systematic nomenclature are difficult. Even the broad division into sedimentary, igneous and metaphoric is by no means clear cut, but is usually helpful, if occasionally confusing.

4.3.1 Sediments

Sedimentary rocks are mostly derived from the past erosion of other rocks and deposition of their fragments and particles, often on the sea bed, followed by consolidation.

Most sediments have been transported by water, but many surface deposits, particularly in the higher latitudes, have been carried by ice, and deposited unsorted as boulder clay or till, or as fluvio-glacial sands and gravels. The terms 'boulder clay' and 'till' for unsorted glacial deposits are largely interchangeable, but the latter is certainly preferable if boulders are absent. Sediments include also organic deposits of vegetable or animal origin, and salt beds resulting from evaporation of lakes and lagoons. When not fully consolidated and cemented they constitute the soils of the soft ground tunneller—gravels, sands, silts, clays—in which particle sizes and water content are of great significance. In consolidated form arenaceous rocks such as sandstone are constituted principally of quartz particles and are correspondingly hard, but may be weakly cemented. Argillaceous rocks are based on clay minerals and are fine grained and soft, forming mudstones and shales.

Limestones are largely of marine organic origin with a high proportion of calcium carbonate; they are typically fine grained and well cemented, but with a well developed joint structure. Their special characteristic of solubility in acidic water may result in enlargement of joints and creation of caverns traversed by underground streams and rivers.

Coal seams are organic strata, originating from swamp vegetation. A particular hazard from coal and other carbonaceous strata is the possible generation of methane gas. Another, man-made, hazard to tunnelling in coal fields is the presence of old worked out mines, possibly unrecorded.

Age of rocks is not of direct importance in tunnelling, but it does give some idea of their probable degree of consolidation and stability. Also, where strata of different periods are found the boundary between them constitutes an unconformity and is likely to mark an important change of conditions and possible plane of weakness. Fig. 4.2 shows some of these features in a section of Tyne Tunnel.

4.3.2 Igneous rocks

Igneous rocks, including volcanic and plutonic rocks, are probably the primary source of the earth's continental crust. Volcanic rocks include lava flows, which have been ejected in a molten state and spread over the existing surface, solidifying by rapid cooling and forming a fine grained crystalline rock such as basalt. They usually fit normally into the stratified pattern of sediments. Sills are intrusive lavas injected beneath the surface in more or less horizontal sheets between pre-existing sedimentary strata. Dykes are typically vertical sheets intruded into joints and fissures; they have sometimes formed feeders for lavas welling up to the surface. Sills and dykes are formed of relatively homogenous crystalline rocks of coarser texture than surface lavas because of a slower rate of cooling. In areas of volcanic activity ash beds, tuffs and ignimbrites may occur in considerable thickness and in various states of consolidation.

Plutonic rocks are formed in large masses at greater depth, not necessarily in a fully molten state, but they are of coarsely crystalline texture. Granite is the typical plutonic rock, but may also be the end product of metamorphic action on sediments.

Many of these igneous rocks are normally dense and impermeable but may be decomposed due to deep weathering at surfaces and at joints. They are further classified according to their chemical composition and crystalline structure. The content of silica, in the form of quartz, is used to divide them into 'acid' and 'basic', with further subdivision according to their feldspar content. The typical acid rocks are light in weight and in colour ranging from rhyolite lava to granite, while the basic rocks are dark and heavy like basalt, dolerite and gablro.

4.3.3 Metamorphic rocks

Metamorphic rocks are in origin sedimentary or igneous rocks which have been substantially altered by heat, pressure or chemical action. At one end of the range they grade into sedimentaries subjected to heavy stresses, and at the other into plutonic rocks. Shearing forces acting on rocks are likely to produce a foliated or laminated structure, as in gneiss and schist, or cleavage as in slate. Marble is a metamorphic rock generated from limestone by recrystallisation at high temperature. Schist derives from clays and shales, and quartzite from sandstones.

In general, metamorphic rocks are ancient and of massive structure, but with

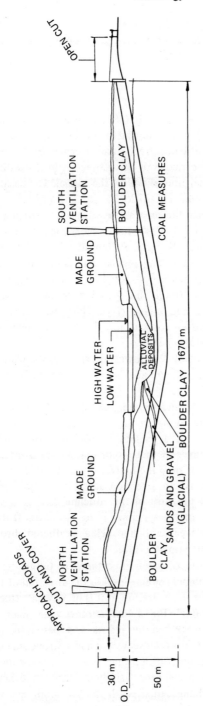

RATIO OF SCALES – H:V 1:3

Fig. 4.2 – Tyne Tunnel – longitudinal section.

planes of weakness which may produce unexpected difficulties. In the less ancient mountain belts formations which are not fully consolidated may become unstable when disturbed by excavation.

4.4 STRATIFICATION

The arrangement of strata may vary from a simple succession of nearly horizontal beds to a complex pattern of folding and faulting inclined at any angle. The strata may include deposits of very local occurrence and irregular thickness. What is required from the survey is a knowledge of the geometry, in three dimensions, of each bed everywhere in the neighbourhood of the tunnel. The basic elements in the description, after the lithology, are thickness, dip, and strike.

The thickness may vary over the area surveyed, more particularly where recent unconsolidated deposits fill hollows in an older land surface, or have been subjected to surface erosion. Superficial deposits of glacial and fluvio-glacial origin may not be stratified in any regular way, and may have been so disturbed and eroded during a period of periglacial weathering as to defy prediction. Deposits, in addition to tills and gravel beds, include such forms as moraines, esker ridges, kames, and varved clays. Older beds may also have been thinned out locally by erosion or cut into by buried stream channels. Lateral variation in lithology may also occur.

Dip and strike define respectively the downward slope of the plane of a bed and the orientation of a level line at right angles to maximum slope and lying in the plane of the bed.

In undisturbed beds, which are uncommon, the dip will be zero. The angle of dip measures the local tilting to which the strata have been subjected by earth movement, usually as part of a pattern of folding into anticlines and synclines over a wide area. The axes of these folds may be horizontal, lying in the same direction as the strike, or may themselves dip more or less steeply and produce a more complex geometry. In a section across a simple fold the dip varies continuously from zero at the crest of the anticline through a maximum slope back to zero at the trough of the syncline. These structural features may be reversed topographically, the weak anticline being weathered into a valley and the strong syncline left standing up as a ridge.

The dip and its variations are of great importance in tunnelling through stratified deposits (Fig. 4.4). A tunnel running level in the direction of the strike may be expected to remain in the same stratum, if it is of adequate thickness. If the dip across the line of the tunnel (Fig. 4.4(a)) is other than small, the roof will tend to break away on any weak bedding planes, and there is likely to be a weak side where rock slides down into the excavation and a strong side opposite, but this may be modified by patterns of jointing. If a tunnel runs level in the direction of the dip (Fig. 4.4(b)) it will progressively cut through higher beds of the succession with consequent changes of strength and behaviour; any weakness

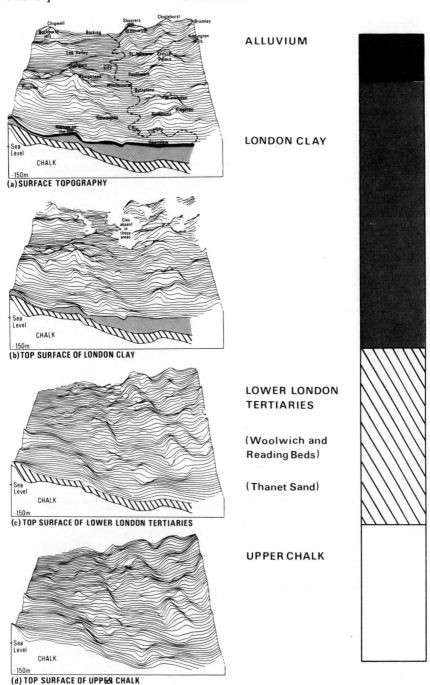

Fig. 4.3 – Geology of London Basin.

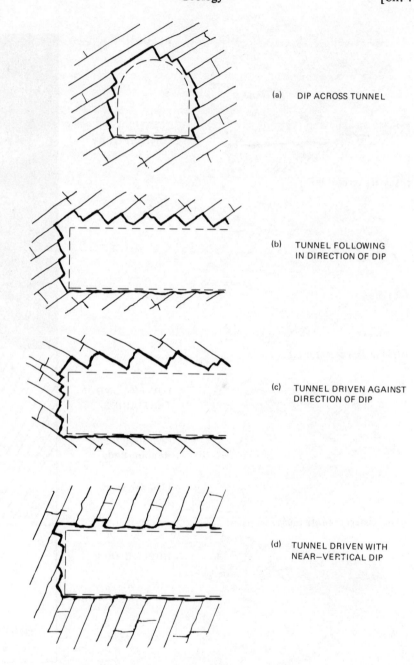

(a) DIP ACROSS TUNNEL

(b) TUNNEL FOLLOWING IN DIRECTION OF DIP

(c) TUNNEL DRIVEN AGAINST DIRECTION OF DIP

(d) TUNNEL DRIVEN WITH NEAR–VERTICAL DIP

Fig. 4.4 – Dip of strata in relation to tunnel excavation.

in shear on the bedding planes will result in weakness of the roof as support is cut away at the face. Conversely, a tunnel driven against the direction of dip (Fig. 4.4(c)) will progress into lower beds and is less likely to be weak near the face.

If the dip is nearly vertical the tunnelling conditions are very different according to whether the tunnel is driven along the line of the strike or at right angles to it. If along the strike, the same beds and bedding planes will persist, and any weakness in shear or any water-bearing stratum may make support very difficult. If the tunnel is along the line of the dip (Fig 4.4(d)) it will cut across the succession of beds at right angles and each bed which is itself competent can function as a transverse arch; a weak bed is only likely to require support while its thickness is penetrated.

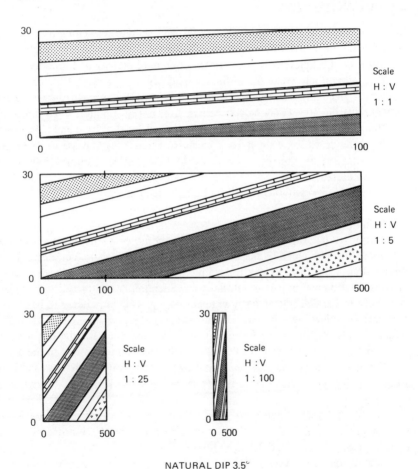

NATURAL DIP 3.5°

Fig. 4.5 – Scale distortion in sections.

In anticlinal structures the rock tends to be more lightly stressed as compared with compression in synclines. Water in fissures therefore may drain more freely from anticlines and accumulate in synclines.

The importance of dip and its variations will be apparent. It may be noted, however, that longitudinal sections of tunnels can be misleading where they are drawn to such different horizontal and vertical scales that they give the impression that all the strata are nearly vertical. For example, with scales in the ratio of 1 : 100, a dip of 5° is exaggerated to 83°. Such a section, therefore, does not give a clear picture of the actual transitions as encountered in driving the tunnel, although showing correctly the successive strata.

4.5 DISCONTINUITIES

In almost any uniform continuous ground, appropriate tunnelling methods can be devised. The more difficult problems arise from discontinuities, particularly if unforeseen. Complete change of construction method may be found necessary.

Bed joints imply some discontinuity, and often a change of lithology which may be of great importance. A change in the crown of a tunnel from sound clay to water-bearing gravel may be very serious; it is not uncommon when tunnelling near the upper surface of London Clay. A 'mixed face' of hard and soft ground can be difficult to excavate if rock excavation requiring the use of explosives causes instability in the soft ground. In the first Blackwall tunnel (1897) rock was encountered in the invert and severe damage to the bottom section of the shield resulted, necessitating replacement in difficult conditions.

4.5.1 Joints

In addition to bedding planes, which arise from minor or major discontinuities of deposition and separate the individual strata, sets of joints perpendicular to the bedding occur in most strata, usually attributable to shrinkage in the process of consolidation and in part to tensile stress. Such joints are approximately plane and occur in parallel sets at fairly regular spacing. There are likely to be at least two sets of joints, which may be at right angles, separating the rock into a mosaic of blocks. In folded sedimentaries it is most usual for the major set of joints to run parallel with the strike, with minor cross joints parallel to the dip.

The spacing of joints may be several metres in massive sandstones and limestones, or may be so small as to fragment the rock and leave it with very little cohesion.

In volcanic rocks there may similarly be sets of shrinkage joints in the vertical plane, or at right angles to the surface of cooling. Massive columnar structures may develop in dolerite and basalt.

In addition to the vertical joints flow layering, or flow foliation, commonly appears in volcanic rocks, and roughly horizontal sheeting joints develop in granites where vertical loading has been relieved by erosion.

Joints may be minimal cracks or relatively open. They may be filled with clay or with crystalline deposits. They provide a network of fissures through which water, or other fluid can travel, contributing to the weathering and erosion of the rock mass. The jointing is of major importance in assessment of the rock quality in relation to excavation and support.

4.5.2 Unconformities

An unconformity is a major structural feature where the sequence of strata is interrupted by an old surface of erosion, which cuts across tilted bedding planes and is buried by a new succession of conformable strata. Dip and strike are likely to change abruptly at this surface, with a complete change of lithology.

The unconformity may be marked by deep weathering of the underlying strata.

4.5.3 Faults

A fault is a fracture surface where relative shearing movement of the adjoining masses of rock has occurred. The extent of movement—the 'throw'—may be very large, or may be so small as to differ little from a joint. Most ordinary faults have ceased movement and are stable, but some major faults, particularly in earthquake zones, remain potentially active. The fault surface is ordinarily nearly plane; its orientation is defined by the dip (or by the complementary angle 'hade') and strike.

There are three typical patterns of fault movement: (1) A normal fault lies in a steeply inclined plane on one side of which the strata have dropped down relative to the identical strata on the upper side. (2) The converse is a thrust fault in which compressive forces have caused movement up the dip slope, which is usually less steep than in a normal fault. With a very low angle the fault becomes an overthrust. (3) The third type is the wrench fault in which the plane is near the vertical and the movement is mainly horizontal to left or right along the strike.

Overthrust faults are particularly associated with the process of mountain building and may mark sliding movements of many kilometres over wide fronts. They are likely to be associated with intense folding and subsidiary faulting and with metamorphism of the strata involved. The classic example is the Moine thrust in the north-west Highlands of Scotland. Such faults also occur on a spectacular scale in the Alps, Himalayas and other mountain ranges. Groups of smaller and steeper reverse faults are often associated with overthrusts.

Wrench faults also may be on a very large scale. One of the best known is the San Andreas fault in California which is over 1200 km in length. Movement totalling 600 km in the course of 140 million years has been identified, and currently the rate has accelerated to about 50 mm per year. A similar fault known as the 'Alpine Fault' in New Zealand extends from SW to NE over most of the length of both islands. Movement has been, and is, at similar rates. Movement at these faults is associated with earthquakes.

A minor fault may have sheared cleanly through the rock, producing polished faces with slickenside, but nearly all faults are accompanied by some degree of distortion and shattering of the adjoining rock in the shear zone, where broken rock may be reduced to rock flour or replaced by clay. The shattered rock is fault breccia and the fine and soft filling is known as 'gouge'. These shear zones provide difficult or hazardous tunnelling conditions. The fragmented rock is difficult to support, and unpredictable in its incidence; gouge may be washed out by water with loss of support to the adjacent blocks and danger of flooding, and clay may provide lubrication in the joints of the rock mass, and consequent instability.

The orientation of a fault and the width of its shear zone in relation to a tunnel being driven through it are obviously very important as are the magnitude and direction of residual stresses in the zone and the presence of water and the extent of its reservoir. A narrow vertical fault through which the tunnel crosses at right angles is likely to be much less difficult than one oblique to the tunnel or with a wide shatter zone. Undersea faults encountered by the Japanese Seikan tunnel have proved very difficult, as described in Volume 1.

4.5.4 Dykes and sills

Dykes and sills have already been referred to above. Dykes may be encountered as vertical walls of intruded igneous rock cutting across sedimentary beds, having followed and enlarged the course of a joint plane or a fault plane. Sills are horizontal intrusions which have followed a bedding plane, forcing apart the strata.

4.6 WATER

Site investigation of surface water and ground water has been discussed in Volume 1, where reference has been made to the danger of inundation. The hazards from internal flooding require some further consideration.

In subaqueous tunnels the major danger is almost always that of inundation by failure of the ground at or immediately behind the face before construction of the lining is complete. The source of water from sea or river, in the event of a breach at the face, is unlimited. Any flow of water, unless immediately stemmed, tends to erode a channel of increasing width and accordingly itself to increase rapidly, although it may eventually become choked with material carried in. Safety precautions and remedial action, apart from the use of a shield, close timbering and use of compressed air, may comprise reduction of permeability and strengthening of the ground by grout injection or freezing, or the use of a clay blanket in the river bed.

In mountain tunnels there are also special risks. Springs fed from fissures may discharge at high pressure and high temperature but the rate of flow is likely to diminish with time as the immediate reservoir becomes exhausted. Such inflows of water, however, may also carry with them solids, washing out the material filling joints, or may carry in sand from a weakly cemented stratum which is thereby disrupted.

In all tunnels the problem of an inflow of water has got to be considered in terms of the extent of the reservoir feeding it, and its erosion and transport of solids.

At very high pressures the stressing of the lining in the finished tunnel by the build-up of full hydrostatic pressure has to be examined. Where acceptable seepage with appropriate provision of drains and pumps allows external pressure to be limited, and in part controlled, this may be preferred to a fully watertight lining. In the case of hydroelectric power tunnels this is unlikely to be acceptable and the lining must be designed for both the alternative cases of high internal pressure in working conditions and high external pore pressure when the system is drained down.

Considerations of watertightness are also of major importance in sewers to ensure that the ground is not polluted by leakage outwards, and in water supply tunnels where contamination by inward leakage must be prevented.

4.6.1 Ground water

Ground water originates as rainfall or snow by soaking into the subsoil and bedrock where it becomes particularly relevant to tunnelling. Its absorption, movement, storage and discharge are regulated by the porosity and permeability of the soils and rocks, and by their geological structure and by the level of the water table. The water passes down through, or is retained in voids in the ground. These voids range from subcapillary dimensions in clay and fine grained sediments, through coarser sands, gravels and conglomerates and open joints and fissures to the extreme of vast caverns in limestone.

The water table is a critical level in the flow pattern. It is the buried surface below which the ground is saturated, and usually lies at some depth below the surface, varying from time to time with rainfall. In boreholes it is displayed as the standing water level, and near lakes and rivers it is nearly at the free water level, rising gently as distance from the open water increases. It marks the surface of a hidden reservoir of water, discharging through springs wherever the water table is above ground level, and sometimes through submerged springs in a river bed or even under the sea.

Above the water table soils and rocks are only partially saturated, and any free water moves downwards, at the rate that permeability allows, until it reaches the water table where it replenishes the reservoir and may raise its surface level. All pore water is not free, much being retained in fine grained material by capillary action. In addition there is adsorbed water bound to particles by molecular action. This adsorbed water is fundamental to the particulate structure and characteristics of a clay and provides shear strength in a way that capillary water does not; it is not readily displaced by consolidation or otherwise.

Below the water table movement through the saturated ground is necessarily much slower, being regulated by permeability and hydraulic gradients to the eventual points of discharge. As already noted, there may be more than one water table level because of the confining effect of impermeable strata.

Porosity measures the voids in rock or soil as a percentage of the total volume and is therefore substantially less than unity; it is not to be confused with 'voids ratio', which is the volume ratio voids : solids. Porosity is a characteristic of a laboratory specimen of rock and does not take account of the joints and fissures and other cavities; with the inclusion of these the mass porosity may be much higher.

Permeability measures the rate at which water can pass through the rock. In particular the coefficient of permeability, k, is, for present purposes, defined by the equation $q = ki$, where q is the rate of flow of water across a unit area and i is the hydraulic gradient in the direction of flow. As with the porosity the mass permeability utilising the joint structure may be much greater than that of an unfissured specimen. The mass permeability may also be strongly directional, following the dominant joint pattern.

Aquifer is the term used for a stratum of permeable rock which may hold water and through which water may pass.

4.6.2 Water in soft ground

The importance of pore water in soft ground lies in its action and effects on the soil. This indeed provides much of the subject matter of soil mechanics. In granular soils and silts the advantages of cohesion provided by dampness, and the disadvantages of either saturation or drying out which result in the loss of cohesion are further discussed below. The effect of any concentration of flow through such soils is to wash out the fines and set the loosened sand and gravel in motion.

In clays and in silts pore water content and pressure are determining factors in the shear strength and plasticity of the soil. Water content is not readily altered in clay, and residual pore water pressures may cause swelling of the clay, creating active pressures on support structures.

Impermeable layers have already been mentioned as giving rise to perched water tables. They may be of variable extent and thickness forming lenses which make further difficulties in the treatment of ground by grouting. In subaqueous soft ground tunnelling the face is extremely vulnerable to the action of water and the danger of inundation. It was to meet these conditions that the Greathead shield was devised, with supplementary equipment such as the hood and face rams. It is also for such ground that compressed air working is adopted, and ground treatment by grouting to reduce permeability and provide greater cohesive strength.

4.6.3 Chemical action

One aspect of ground water which may be of importance is chemical activity. Acidity in the ground water and sulphate content may attack cement used in grouting and pointing or concrete linings. The problem of attack on cement is principally one of flowing water which can carry away the reaction products and expose new surfaces to attack. Strong acid action has been observed in a cast iron lining in London Clay, probably attributable to oxidation of iron pyrites with resulting sulphate formation. Swelling ground attributable to chemical reaction is further referred to below.

4.6.4 Flotation

Another aspect of ground water in shallow tunnels is the possibility of flotation. An immersed tube tunnel, or one rising from a subaqueous crossing through soft unconsolidated ground, may be subject to nett uplift, requiring anchorage or ballasting against uplift.

4.7 ROCK STRESSES

In the normal configuration of stress in undisturbed rock the major principal stress is vertical and approximates to the direct pressure of the overburden. Accompanying horizontal stresses are about one quarter to one half as great. The significance of horizontal stress has been discussed in Volume 1. Abnormal stresses may be present in the rock or may be developed as a consequence of tunnelling operations. The principal phenomena to be considered are:

1. Stress exceeding rock strength.
2. High values of horizontal stress, or inclined stress, which have resulted from tectonic activity in the past during folding and faulting of the strata.
3. Inclined and irregular stresses of topographic origin, as in the steep side slopes of a valley.
4. Squeezing rock, where the disturbance resulting from excavation allows a stratum to yield in a plastic manner.
5. Swelling rock where physical or chemical action causes a stratum to swell and to impose heavy loading on tunnel support structures.

1. Direct stress of normal pattern may exceed locally or generally the unsupported strength of the rock during excavation. Rock bursts, popping rock or spalling may result. It is a familiar phenomenon below 600 m in deep gold mines where controlled de-stressing can alleviate it. It was noted in the chalk excavation preliminary to Channel tunnel construction in the walls of an arched heading with rib supports. It did not appear for some time after excavation and could be controlled by providing lateral support immediately after exposure. In a New York water tunnel under the East River at a depth of 130 m below ground popping rock was common in dolomite and metamorphic

rocks of Cambrian and pre-Cambrian age. Rock bolting was ineffective but steel ribs with timber lagging protected the work satisfactorily.

2. Abnormal horizontal stress may be difficult to ascertain from preliminary investigations, but the possibility of occurrence may be predictable if the geological history of the zone is known.

3. In hilly or mountainous country sloping ground is likely to have special characteristics as one principal stress may be inclined in the direction of the slope, but the pattern will depend on the relationship between the dip of the strata and the slope of the hillside. If both dip in the same direction, a tendency to slip is probable while if the dip is against the slope the ground will tend to be more stable.

The slope surface may be unstable, or of low stability against sliding if it consists of scree and hill wash above rock or if the rock is deeply weathered. These factors become of great importance in tunnelling either for construction of a portal entering the hill slope at right angles, or for a tunnel running nearly parallel to the slope.

4. Squeezing rock may develop in mountain areas where heavy stresses deform weak rock and cause it to yield and flow. When excavation exposes the squeezing rock it is extruded into the tunnel without visible fracture and without change in water content. The rock behaves like plastic clay, the rate of squeeze depending on the overstress, measured by the ratio of overburden pressure and shear strength.

The pressure on supports may reach very high values.

Squeezing rock can develop from metamorphism of shales or schists by the chemical action of deep weathering breaking down the original structure.

5. Swelling ground is attributable to change in water content causing expansion. One form is overconsolidated clay in which pore water content has been determined by former overburden pressures from strata long since removed by erosion. London Clay is a well known example. Excavation at a tunnel face releases pore water pressures and slow migration of water from the body of the clay towards the face causes swelling at the face.

Other types of swelling ground include beds of anhydrite ($CaSO_4$), in which access of water results in chemical change to gypsum with an increase of volume.

Magnetic iron pyrites, pyrhotite, decomposing to sulphuric acid and ferrous sulphate when exposed to oxygen and water, can result in swelling accompanied by corrosion problems.

4.8 GEOTECHNOLOGY

In order to make the best use of the information embodied in the geological survey and to assess the tunnelling conditions, it must be further analysed in terms of the mechanical behaviour of the ground during and after excavation.

As in all civil engineering, the old practical and empirical assessments of soils and rock have been largely replaced by the quantitative analytical methods, first of soil mechanics and then also of rock mechanics. In the application of these methods it remains essential to exercise at all stages engineering judgement based on experience. The siting of boreholes, the selection of samples for testing and the applicability of those test results all have to be considered and kept under review in the light of the actual ground encountered and its response to excavation.

It may be worth while to quote Terzaghi's words from his *Theoretical Soil Mechanics* (1943): 'The prospects of computing accurately the effect of a change in the conditions of loading or of drainage on the soil in advance of construction are usually slight—particularly (where) the action of water is involved. The action often depends on minor details of stratification which cannot be detected by test borings.'

4.8.1 Soil description
Soil mechanics is particularly related to the interaction of fine particles and water in the soils encountered in soft ground tunnelling, namely: gravel, sand, silt and clay, and their modification with changes of water content and under stress. In ground of this type jointing patterns are unlikely to be relevant,

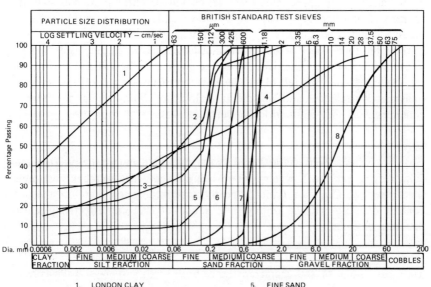

PARTICLE SIZE ANALYSIS — TYPICAL GRADING CURVES

Fig. 4.6 — Soil grading curves.

except possibly in very stiff fissured clay. The presence and pressure of water, adsorbed, capillary, in pores and in fissures, can be of dominant importance as has been discussed above and in other chapters.

The primary mechanical classification of soils is by particle size distribution, followed by mineral composition and cohesive and plastic properties. The results of sieve analysis are best displayed by plotting a curve which shows the percentage of the sample passing each standard sieve against a logarithmic scale of particle diameter. Figure 4.6 shows some typical curves and shows also the size limits and the terminology. For the finer part of the range, below about 0.06 mm, sieves cease to be usable and sedimentation rates or other methods of estimating size are adopted. This is size grading only and it must be recognised that, for example, clay size particles do not necessarily have the characteristics of clay if their mineral constitution differs: finely ground quartz, as in rock flour, will behave quite differently from a true clay.

From sieve analysis a soil is classified descriptively by the name of its principal constituent, qualified by naming any other substantial fraction as, for example, clayey sand, silty clay. These soils fall into three main subdivisions: granular soils, silts, clays.

Granular soils
Such soils lack any appreciable fine fraction and have little or no cohesion when submerged or when dry. They are therefore liable to run in either such state unless adequately contained and supported. When damp they have some cohesion attributable to capillary action of the moisture. This can be of considerable help in tunnelling, but its value depends on the grading of the soil and presence of a clay fraction and its reliability depends on maintenance of the necessary moisture. Water levels and any drainage flows, or drying out as by escape of compressed air, may be of critical importance.

Clays
Clays are cohesive soils characterised by shear strength and plasticity. They range from soft to very stiff, water content being a significant factor in determining shear strength. Clays are typically nearly impermeable and therefore the water content does not change readily, and a bed of clay can effectively seal off intrusion of water.

Silts
They have some of the characteristics of both sands and clays. Their cohesion is dependent on water content remaining between narrow limits. Their permeability is such that they may become saturated, or dry out, fairly rapidly, losing all cohesion and running into an excavation unless retained by very close-fitting timbering. On the other hand, their permeability is too low to allow grout injection except with extremely penetrative fluids.

4.8.2 Soil properties

In terms of soil mechanics the principal properties of soils to be obtained from samples include:

> Particle size grading, as already discussed
> Mineral composition
> Density
> Porosity
> Moisture content, from an undisturbed sample if possible
> Shear strength, in clays
> Angle of internal friction in granular soils
> Plasticity—plastic limits and liquid limits in clays and silts

Other properties, important in most circumstances, must be determined by *in situ* measurements. These include:

> Permeability
> Pore water pressure
> Flow of ground water
> State of compaction
> Fissuring

4.8.3 Quality of rock

For tunnelling in rock a similar list of properties can be given, again with the obvious proviso that not every sample needs to have all these tests made. What will be sufficient is very much a matter of judgement, influenced by time and cost. The information obtainable from core samples includes:

> Lithology
> Density
> Porosity
> Water content
> Strength, including crushing and tensile strength and shear
>   or any identifiable plane of weakness
> Abrasiveness, in relation to cutting by drill bits or by
>   machine cutters
> Rock Quality Designation (RQD), being the percentage
>   of core borings recovered in lengths not less than
>   100 mm

Site investigations are necessary to ascertain:

> Permeability
> Pore water pressure and flow of water
> Joint and fissure pattern

Attempts have been made in recent years to establish quantitative assessment of the tunnelling qualities of the rock mass. One that is widely used is the 'rock mass quality index' of the Norwegian Geotechnical Institute, which provides an example of the complexities necessarily involved even although it still includes a large empirical element. Experience and sound judgement remain essential in evaluating the index and interpreting the result.

The index, Q, is compounded from numbers selected to represent RQD (as defined above), number of sets of joints, joint roughness, joint alteration, water pressure in joints, and a stress reduction factor allowing for such features as zones of weakness, high rock stress, or squeezing or swelling rock.

Q may lie in the range from .001 for the worst tunnelling conditions in squeezing rock up to 1000 for massive rock requiring no support. This range of values is not very convenient for description and ready appreciation, and might well be translated to a logarithmic scale, perhaps from zero to + 60.

It may be noted that, despite the complexities and the judgement required in estimating most of the components of the index, there remain other properties of the rock to be considered, including compressive strength, orientation of bedding planes and joints in relation to tunnel excavation, hardness, abrasiveness, and fragmentation in relation to the use of cutting tools, blasting methods and spoil handling.

4.9 EARTHQUAKES

Earthquakes are hazardous to all surface and underground structures. They arise typically from sudden release of stresses at depths ranging up to several hundred kilometres, often by movement in the plane of a fault. They manifest themselves, in diminishing degree as distance from the focus increases, by horizontal and vertical vibrations of the ground, by vertical or horizontal displacements across a fault plane, by opening of fissures and by consolidation and slipping of unstable ground.

Earthquakes may occur almost anywhere but the major zones of activity are the margins of the Pacific Ocean and a belt from the Mediterranean to the Himalayas and the East Indies.

Substantial surface structures are principally damaged by the horizontal component of acceleration in the vibrations at the surface, particularly in unconsolidated strata. No structure can, of course, resist a shearing movement in the plane of a fault if the structure is continuous across that plane. Tunnels in solid ground are unlikely to be severely damaged by vibrations unless the ground itself is sheared. There is, however, obvious danger of landslips at tunnel portals or in the sides of a mountain valley or gorge.

The B.A.R.T. tube 6 km long crossing San Francisco Bay was provided at both ends with joints designed to allow for earthquake movements. These joints allow for a longitudinal relative movement of 38 mm in either direction and a transverse movement of 100 mm in any direction.

BIBLIOGRAPHY

The subjects of geology, soil mechanics, rock mechanics, hydrology and other specialisations have vast bibliographies of their own ranging far beyond the aspects considered here. A brief list only, of items of particular relevance, is appended.

Terzaghi, K., *Theoretical Soil Mechanics*, J. Wiley, 1943.

Glossop, R. and Skempton, A. W., Particle size in silt and sand, *J. Instn. Civ. Engrs*, 1945, **25** (Dec.).

Terzaghi, K., Rock defects and loads on tunnel supports: Introduction to tunnel geology in Proctor, R. V. and White, T. L., *Rock tunneling with steel supports*, Commercial Shearing and Stamping Co., Ohio, 1946, 2nd edn. 1968.

Todd, D. K., *Groundwater hydrology*, J. Wiley, 1959; 2nd edn., 1980.

Chow, V. T. (Ed.), *Handbook of applied hydrology*, McGraw Hill, 1965.

Terzaghi, K. and Peck, R. B., *Soil Mechanics in engineering practice*, J. Wiley, 2nd edn., 1967.

Stagg, K. G. and Zienkiewicz, O. C., *Rock mechanics in engineering practice*, J. Wiley, 1968.

Wahlstrom, E. E., *Tunneling in rock*, Elsevier, 1973.

Barton, N. *et al.*, Engineering classification of rock masses for the design of tunnel support, *Rock Mechanics*, 1974, **6**, 189–236.

Blyth, F. G. H. and de Freitas, M. H., *A geology for engineers*, E. Arnold, 6th edn., 1974.

Blyth, F. G. H., *Geological maps and their interpretation*, E. Arnold, 2nd edn., 1976.

Manual of applied geology for engineers, Institution of Civil Engineers and H.M.S.O., 1976.

Jaeger, J. C. and Cook, N. G. W., *Fundamentals of rock mechanics*, Chapman and Hall, 2nd edn., 1976.

Cedergren, H. R., *Seepage, drainage, and flow nets*, J. Wiley, 2nd edn., 1977.

Holmes, A., *Principles of physical geology*, Nelson, 3rd edn., 1978.

Jaeger, C., *Rock mechanics and engineering*, Cambridge U.P., 2nd edn., 1979.

West, G. and Lake, L. M., *Site investigation for tunnels*, Int. J. Rock Mech. Min. Sci., 1981, **18**(5).

5

Ground Treatment

5.1 OBJECTIVES

Preparatory treatment of the ground to modify its behaviour during excavation for tunnelling can take various forms. The principal objectives are control of water and strengthening of cohesive forces, and the means include: (1) injection of grouts into pores and fissures; (2) freezing; and (3) lowering of ground water levels by drainage.

It is in tunnels and shafts below the water table, through granular soils which are permeable to water and lacking cohesive strength, that the usefulness of such methods is greatest, but they may also be employed to control water inflows and stabilise the ground in fissured and jointed rock or even clay. Compressed air working, described in Volume 1, is a very special technique for control or exclusion of water, and its application may be reinforced if necessary by most of the methods of ground treatment described herein.

Normally ground treatment methods are most effectively applied from the surface in advance of tunnel excavation, but they can also be applied from within a tunnel although such operations are almost inevitably in cramped conditions and greatly delay tunnelling progress, unless executed separately from a pilot or parallel tunnel.

When the Thames tunnel of the elder Brunel suffered a collapse, and he received innumerable suggestions from amateurs, he wrote: 'In every case they made the ground to suit their plan and not their plan to suit the ground'. In some cases, nowadays, ground treatment techniques can do just that; running sand can be cemented into a sandstone, wet silt can be frozen to a coherent solid.

5.2 GROUTING

The term 'grouting' describes the injection under pressure of a fluid grout, which may be a solution or a particle suspension, to occupy the pores, or the voids in the ground, or between a structure and the ground. Where the ground is water-bearing the grout is designed to expel and to replace the water, or, in the absence of water to fill the voids. When thus injected the grout reacts, physically or chemically, and may set to form a gel or a solid, reducing the porosity and usually increasing the cohesive strength of the ground, possibly to a very substantial degree.

An increasingly wide range of materials and of grouting techniques and equipment are available to suit particular requirements.

The two principal objectives are, as already stated:

1. to reduce the permeability of the ground, and
2. to strengthen and stabilise the ground.

In tunnel and shaft construction both are usually required. When working in compressed air the reduction in air loss may be just as important as the reduction in water inflow. Suitable filling material in the voids will contribute an increase in the shear and compressive strength of the soil at the same time as reducing permeability, but the appropriate materials and technique of application to suit the ground must be employed. If the volume to be treated contains a variety of soils, more than one type of grout may be required and even in a single stratum successive injections of finer grouts may be necessary.

5.2.1 Grout types
Grouts may be classified very broadly into:

1. Particle suspensions. Portland cement in water is the most familiar: this is only applicable where pore dimensions are sufficiently large to admit the cement particles. Bentonite or other clay suspensions are finer grained and therefore penetrate more freely.
2. Chemical grouts. These are aqueous solutions constituting pure liquids. Their penetration properties depend on their viscosity, which may increase with time after a physical or chemical action of setting to a gel is initiated. A 'one-shot' time-dependent fluid may be used, or a 'two-shot' process in which two different materials are injected in succession and react in the ground.
3. Others, such as bitumen emulsions, asphalt, etc., are not much used in tunnelling.

The choice and design of the most appropriate material and technique and the sequence and pattern of injection can only be made after a comprehensive site investigation with particular emphasis on permeability and sieve analysis. Specialists in grouting are likely to be required for all but the simplest cases, preferably with an ability to modify their techniques as experience is gained in the response of the particular ground.

It is seldom that grouting is completely effective, because grout may fail to penetrate equally all of the volume and some residual permeability is likely. A balance has to be struck between the cost in time and money of increasingly elaborate detailed treatment and acceptance of a sufficient improvement of ground stability and ground water control to make possible satisfactory permanent construction.

5.2.2 History of grouting

Ground improvement by grouting for civil engineering works was introduced early in the 19th century largely for marine and river works. Various materials were tried such as clays, pozzolana mortar, lime mortar or bitumen emulsions. In the last quarter of the century the use of Portland cement mortar became widespread. The use of grout by Greathead in shield tunnelling to fill voids between lining and excavated ground was an important advance, but is not directly relevant to this chapter, which is concerned with preparatory ground treatment.

In the middle of the 19th century grouting in the U.S.A. was developed in dam construction to strengthen foundations and form cut-off walls in pervious ground.

Chemically reactive grout comprising aluminium sulphate solution and waterglass which combine to form a soft silica gel was first used in 1911 at a colliery near Doncaster. In the well known and widely used Joosten process, patented in 1925, successive injections of sodium silicate and calcium chloride penetrate sands and gravels well and react to bind the sand very effectively.

In the last 30 years a wide range of different grouts have been introduced, and concurrently the necessary high pressure pumps and other injection equipment have been developed. Their applications to tunnelling have been extensive and they have become indispensable in difficult conditions.

The first Dartford tunnel (1963) marked the introduction of ground treatment to control losses of compressed air by injection of a range of grouts, including bentonite/cement, into a bed of open gravel through which the tunnel penetrated under the doubtfully stable river flood banks. The complex pattern of injections from an array of tubes at the surface, and from jetties in the river utilising the 'tube a manchette' (sleeved tube), was effective in limiting air losses.

5.2.3 Principles of grout injection

The principal factors governing the choice of grout mixtures and methods of application are:

1. Ground characteristics
    (a) Permeability of the ground
    (b) Flow of ground water
    (c) Ground water pressure
    (d) Depth and accessibility of zone to be treated

2. Grout properties
   (a) Size of particles in suspension
   (b) Viscosity
   (c) Setting time and rate of setting
   (d) Grouting pressure and injection rate
   (e) Chemical compatability

The overall permeability of the ground derives in part from the pore structure of the soil or rock and in part from the pattern of joints and fissures. What follows relates more particularly to unconsolidated granular soils, which are not characterised by joints or fissures.

Permeability, in this context, is defined by D'Arcy's Law relating to percolation of water through a uniform medium, such as sand: $v = ki$, in which v is the discharge velocity of water (cm/s), i is the hydraulic gradient (No), and k is the coefficient of permeability (cm/s).

While knowledge of k will give an immediate value for passage of water, the penetration of a grout will be restricted by its viscosity and by the action of solids held in suspension.

In soils in which k exceeds 10^{-1} cm/s slurry grouts of cement, or bentonite or other clay can be injected at low pressure. Down to $10^{-2.5}$ cm/s chemical grouts of relatively high viscosity, such as those used in the Joosten process are effective, but where permeability is lower resin grouts are likely to be necessary. The practical limit for these grouts is about 10^{-5} cm/s, which excludes fine silts.

Down to this level most soils can be stabilised and rendered impermeable, but the finer the pore structure the more difficult, slow and costly does effective grouting become. Where the strata are variable the grout injections will initially follow the path of least resistance, and lenses of fine silt or even fine sand may remain untreated and may cause trouble subsequently.

If ground water is flowing through the ground the viscosity and setting time of the grout must be such as to ensure its retention in the zone being treated. The pore pressure in the ground is obviously significant, as the injection pressure must be sufficiently in excess to expel the ground water throughout the desired radius of action.

The depth and accessibility of the treatment zone may range from a shallow bed easily reached from the surface to a site where a tunnel is to be driven below a river bed or under a built-up area, and where a pilot tunnel may be the only means of access. In shallow working vertical drillings can be made at close centres but wider spacings become necessary for both technical and economic reasons as depth increases, and the radius of effective penetration of the grout may become of critical importance. If grouting is to be done radially from within a pilot tunnel, as further discussed below, the problems of balancing spacing, penetration and setting time may be difficult.

Grout formulations are examined in more detail below. Where a particulate suspension is to be used the relationship between grain size of grout particles and pore dimensions in the ground is of critical importance. The thickness of the grout must be adequate to give a set to the appropriate strength in the time available, but not such as to demand excessive injection pressure. In liquid grouts viscosity determines the rate of penetration at a specified working pressure. The process of setting implies an increase of viscosity with time,

Fig. 5.1 – Cement grout in chalk. The grout was exposed during enlargement of a pilot tunnel. The cement had filled the fissures, and lifted the chalk bodily.

and this may be either slow and continuous or more abrupt. Time is therefore of great importance in determining the extent of penetration.

Grouting pressure has to be carefully calculated and controlled. An attempt to accelerate the injection process may simply result in penetration of grout locally along a plane of low resistance and its escape beyond the treatment zone, or the build up of a high hydraulic pressure, disrupting the ground and possibly causing heave at the surface without effectively treating the finer sands. This process can be deliberately employed to stabilise a fissured clay or other impermeable material by establishing a network of fissures filled with a strong grout.

The question of chemical compatability between grout and ground may have to be considered, as for example in ground of high acidity, or alkalinity, or in the presence of sulphates. Some grouts have toxic constituents, the handling and ultimate dispersal of which may have to be very carefully controlled. Any possible dangers of poisoning water in the ground also have to be considered.

Fissure grouting is aimed at the filling of fissures and joints in rock, and possibly in stiff clay, to reduce the intrusion of water and to minimise movement of blocks disturbed by tunnel excavation. The same principles apply but details of application and technique differ. Grouting of shattered fault zones ahead of excavation may be essential to safe construction. Pilot drilling ahead of the face as far as practicable is required to locate major water-bearing fissures and to allow them to be sealed by grouting. Whatever pattern of drilling is adopted for exploration and for treatment, it is impossible to ensure that every fissure and joint is located and treated; a boring may run nearly parallel to the plane of a major fissure and fail to intersect it. In some conditions fissures may be apparently blocked with soft 'gouge' which may be washed out by inflowing water when exposed. A useful treatment may be the deliberate washing out of the gouge by high pressure water, followed by refilling with grout.

5.2.4 Grouting materials
5.2.4.1 Particles in suspension
Cement Ordinary Portland cement is used in gravels and coarse sands, and also in fissure grouting. The average particle size of ordinary commercial Portland cement is 30 μm, but individual particles may be up to 100 μm. It is useful where passages for grout flow are of the order of 10 times the particle size. This corresponds to ground having a permeability greater than $10^{-0.5}$ cm/s.

To obtain maximum penetration by the cement a technique of progressive thickening for successive injections is used. An initial thin mix, such as 100 l water to 50 kg cement, allows the grout to reach and penetrate and seal the finer voids, excess water being expelled into the surrounding ground by maintained pressure. Successive stages with reducing water/cement ratio ensure also the filling of the coarser voids. The procedure ensures maximum strength and minimum permeability. Initial setting time is about one hour, but hardening is a slower process.

This progression with a single relatively cheap material contrasts with the technique of initially using cement grout and subsequently infilling the finer zones with expensive low viscosity grouts.

Rapid hardening cement is more finely ground and the smaller particle size increases penetration, with the possible added advantage of shorter setting time and high early strength.

High alumina cement gives early strength and resists sulphate and acid attack but cannot be used in alkaline conditions. When mixed with ordinary Portland cement there is a very rapid set, useful for plugging leaks or other emergency action.

Supersulphated cement is a finely ground mixture produced from granulated blast furnace slag and ordinary Portland cement with added calcium sulphate. It has good penetration and is highly resistant to sea water, acids and sulphates.

Other cement based mixtures may be adopted for special purposes. Addition of sand will provide a cheaper grout for filling large cavities.

Cement/Clay Mixtures of cement with clays can offer a cheaper filler grout than pure cement. The strength is lower but watertightness need not be impaired. Not all clays are suitable: Kaolinite and Illite are commonly used.

Bentonite is a very special clay, which has thixotropic properties when mixed with water, that is to say it behaves as a fluid while in motion or if agitated, but when stationary forms a gel, with properties of low permeability and some cohesive strength. It may be used alone to reduce permeability in gravels and coarse sands, if substantial strength is not required.

Cement/bentonite mixtures are used in fairly open soils and in fissure grouting. The bentonite improves the flow characteristics and does not significantly reduce strength if it is used sparingly.

Pozzolans Pozzolans are finely divided materials insoluble in water, but chemically reactive, named from a typical volcanic tuff occurring near Naples. They have themselves some cementing properties and can be used as additives to Portland cement. Pulverised fuel ash (PFA) is a waste product of coal-fired power stations, having similar properties, and is readily available. It has largely replaced natural pozzolans. Fly ash, also from power stations, has similar properties and uses.

PFA mixed with about 35% water forms a slurry, and further dilution up to 50% makes it flow freely, but the amount of water may be rather critical for particular pumping and injection requirements. It will react with lime and water to form a stable weak cement, but its particle size is in the silt range and its use in soils is restricted to gravels and coarse sands.

PFA and fly ash can provide economical fillings for large cavities.

PFA/cement in proportions up to 20 : 1 produce a cheap material with good strength characteristics, but large particle size limits its applications.

5.2.4.2 Chemical grouts
Those currently used may be grouped as silicates, lignins and resins.

Silicates The oldest and best known silicate injection technique is the Joosten two-shot process, in which sodium silicate is first injected followed by a solution of calcium chloride, which sets up an immediate chemical reaction resulting in a calcium silicate gel which binds sand into a weak sandstone. The consistency of the gel may not be uniform, being greatly influenced by variations in ground porosity. The process is well tried and is effective if expertly applied, but its use is restricted by the high viscosity of the silicate, because of which satisfactory penetration cannot be expected in gradings finer than medium sand.

The Guttman process is similar but the viscosity of the silicate is reduced by dilution with sodium carbonate, which improves penetration but gives lower gel strength.

One-shot processes have been developed by mixing the silicate prior to injection with appropriate reagents. The setting time for the reaction allows for the injected grout to penetrate sufficiently, but strength may be low and these are mainly used where reduction of permeability is the principal objective.

Examples of these additives are: weak gels—sodium bicarbonate, lime water, sodium aluminate; strong gels—ethyl acetate, amide, citric or oxalic acid plus a salt of aluminium or iron.

Lignins Low viscosity grouts with penetrating power adequate for fine sands can be developed from lignosulphite, a by-product of the wood pulp industry, in combination with bichromates. The mixing of these liquids results in formation of a firm gelatinous mass, with a setting time ranging from 10 mins to 10 hrs, dependent on the concentration of bichromate. The viscosity and final gel strength also vary. Some of the bichromate grouts may be toxic at certain stages and appropriate precautions in handling them are required.

Resins The grouts of greatest penetration are those produced by polymerisation of organic chemicals, including resins, which can be used at permeabilities as low as $k = 10^{-4}$ cm/s. They are mostly proprietary formulations, and are adapted to the particular conditions expected. They include:

AM-9 (American Cyanamid Co.) which is a mixture of acrylamide and a methyl derivative. Its setting time may be affected by the presence of salts in the groundwater and by its pH value.

Resorcinal with formaldehyde (Cementation Co.) which are injected together in aqueous solution, producing a polymerised resin with properties partly elastic and partly plastic.

Geoseal Resin Grouts (Borden Chemical Co.) are based on water-soluble resins in powder form, activated by caustic soda, with various modifications and additions to control viscosity and gel time. The grouts are described as MQ4, MQ5, etc. A two-shot grout is included in the range.

Other materials are sometimes used for special purposes. Bitumen emulsion can be used in medium sands to reduce permeability, but the cohesive strength is low. Hot bitumen can sometimes be injected to seal large cavities in the presence of flowing water.

Polyurethane is an oil based polymer with relative high viscosity.

Cyanloc 62 is a resin grout not water-soluble which forms a very stiff gel when mixed with sodium bisulphite. Being injected as a particle suspension the penetration is limited.

5.2.5 Equipment

The essential plant for ground treatment by grouting comprises:

1. Drilling gear
2. Batching and mixing equipment
3. Pumps
4. Grout injection tubes

The operations may range from relatively simple processes, such as cement or bentonite grouting from the surface. to treatment of varied and difficult ground with a succession of specialised grouts, and injections at high pressures from a confined space in a pilot tunnel. In all cases careful control and accurate records by skilled operators are required, and in the more complex applications a site laboratory and expert supervision by specialists are likely to be essential to a satisfactory outcome. Many of the chemical and resin grouts involve patented processes, and full details of formulation and use may not be published.

Percussion and rotating drills are both used. Casing will be provided or drilling mud used only if necessary to prevent collapse. There are specialised techniques which allow controlled change of direction during drilling.

Batching and mixing plant must be carefully planned and sited. It may be no more than a tank in which to mix cement and water, or a tunnel grout pan employing compressed air for injection, but, with the special chemical grouts, very precise proportioning of the ingredients and accurate timing of the mixing and injection processes may be necessary to ensure the correct viscosity and setting time; the plant may be very elaborate with extensive instrumentation. One important aspect is to ensure that the grout does not set prematurely and solidify in the pumping and injection system.

Pumps must allow control of quantities injected and pressures employed, and are therefore normally of the positive displacement type. With a two-shot system the use of separate pumps to avoid mutual contamination is obvious but also requirements for injection rates and pressures may be different. In some operations the larger voids are first filled with a relatively coarse and cheap material and followed up by infilling with a more penetrative chemical grout.

Grout tubes may be simple open-ended tubes, possibly fitted with an expendable tip to prevent blockage in driving, or perforated tubes which allow grout to be injected over a length may be used. The 'tube-a-manchette' makes possible successive injections of different grouts and choice of the level for injection. Perforations at appropriate intervals are closed by an external elastic sleeve which can be opened by the internal pressure of the grout; the level of injection at any time is regulated by pistons above and below the injection point (see Figure 5.2). In mixed ground initial injection of cement/bentonite in gravel and coarse sand can be followed by chemical and resin grouts in finer sands which the particulate grouts cannot penetrate, and injections can be repeated until the required reduction of permeability is effected.

Spacing of grout tubes normally is within the range 0.5 m to 4 m depending on the nature and extent of the required modification of the ground and the permeability and viscosity.

Where the surface is accessible modern drilling equipment will reach any depths likely to be required for tunnelling and associated shaft sinking.

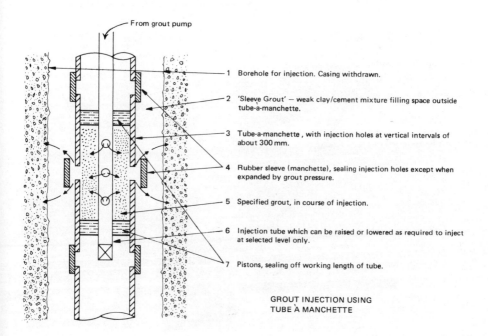

1 Borehole for injection. Casing withdrawn.

2 'Sleeve Grout' — weak clay/cement mixture filling space outside tube-a-manchette.

3 Tube-a-manchette , with injection holes at vertical intervals of about 300 mm.

4 Rubber sleeve (manchette), sealing injection holes except when expanded by grout pressure.

5 Specified grout, in course of injection.

6 Injection tube which can be raised or lowered as required to inject at selected level only.

7 Pistons, sealing off working length of tube.

GROUT INJECTION USING
TUBE A MANCHETTE

Fig. 5.2 — (a) 'Tube-a-manchette'.

Fig. 5.2 – Tube-a-manchette (b) Perforated tube and rubber sleeve.

Fig. 5.2 – Tube-a-manchette (c) Internal piston.

5.2.6 Application

The treatment may be applied to the whole of a permeable or incoherent stratum affecting a tunnel, more particularly the area immediately outside the excavation. There may be an advantage in treating also the ground which is in fact to be excavated, as the added stability of the face will simplify face support, or may allow it to be dispensed with, and control of water will be improved. Selective treatment of critical zones can be economical in time and materials, but more precise control and supervision are necessary.

Access from the surface may be impracticable because of the presence of buildings or other obstacles, or in subaqueous tunnels. There are then the possibilities of treating the ground ahead of the face from the length of tunnel already excavated, or from a pilot tunnel, or a parallel tunnel already completed.

In working ahead through the tunnel face in normal conditions the treatment of ground outside the tunnel must be by a diverging cone of grout pipes. The length which can be treated is then limited largely by the accuracy with which pipes can be driven, and also by the angle of divergence of the pipes. In special cases a sufficient enlargement of the last length of tunnel will allow a ring of parallel pipes to be driven in a cylindrical pattern.

It may not be necessary to treat the whole circumference of the tunnel: a 'hood' over the crown of the excavation may be adequate, or other limited arcs where a thin layer of permeable ground intersects the excavation.

One major disadvantage of working from the tunnel itself is that ground treatment and tunnel excavation cannot proceed simultaneously, the appropriate plant and the specialised labour for the two operations alternating, perhaps on a two-week cycle. The tunnelling is then slow and costly.

Ground treatment in advance from a pilot tunnel has the advantage of continuity of operation and less disruption when the unexpected is encountered. Where a large tunnel is to be built a pilot may be driven inside its area and may be concentric or at high or low level. Grout pipes are driven out through prepared holes in the pilot tunnel lining and must be more or less radial. This imposes two limitations: the length of pipe which can be inserted without a joint, and working space generally, are limited by the diameter of the pilot tunnel, and the radial pattern of pipes means that spacing increases rapidly outwards.

The second Blackwall tunnel provided an interesting example of such ground treatment. The tunnel carries a highway under the tidal River Thames and is cast-iron lined with a diameter of 8.6 m. As much as possible of the tunnel lies within the London Clay, but there is a buried channel in mid-river, cut down in glacial times below the present bed level right through the stratum of clay and subsequently refilled, mainly with sands and gravels. The main tunnel, to be driven with the aid of compressed air and a hooded shield, had to pass through a full face of sand and gravel with a cover of about 6 m. The gravel generally had a permeability of about 10^{-1} cm/s and improvement to an average of 3×10^{-4} cm/s was sought. Two 2 m diameter pilot tunnels were

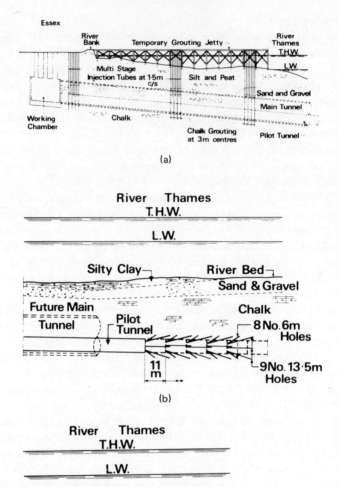

(a)

(b)

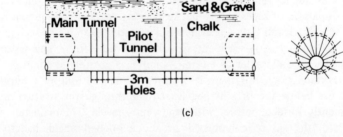

(c)

Fig. 5.3 – Three arrangements for ground treatment below river, for second Dartford Tunnel. (a) Under foreshore, from surface by use of jetty. (b) In a conical array, from face of pilot tunnel, which was constructed in 11 m lengths. (c) Radially, from pilot tunnel preparatory to excavation of main tunnel.

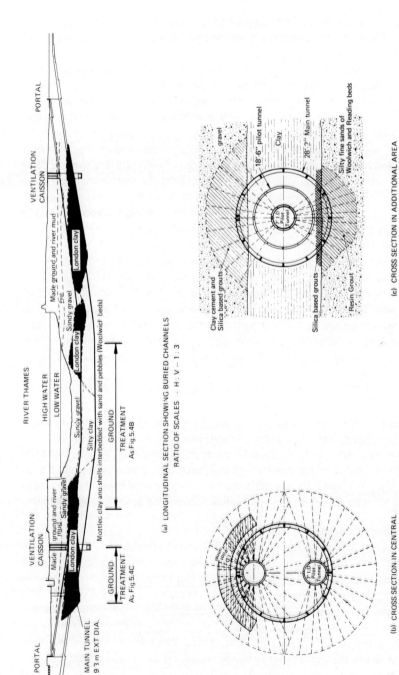

(a) LONGITUDINAL SECTION SHOWING BURIED CHANNELS

RATIO OF SCALES – H : V – 1 : 3

(b) CROSS SECTION IN CENTRAL BURIED CHANNEL

(c) CROSS SECTION IN ADDITIONAL AREA OF TREATMENT AT NORTH END

Fig. 5.4 – Blackwall Tunnel Duplication – ground treatment.

driven in advance as shown in Figure 5.4 to allow the grouting to be carried out. The small diameter was chosen to minimise risks of a blow or inundation during the driving of the pilots under compressed air and with shields.

The whole area surrounding the tunnel was injected with clay/chemical grout effecting the necessary reduction in permeability. In addition, because the working pressure in the main tunnel might have to be raised above the hydrostatic head at the crown, an arched canopy of ground above was further injected with a low viscosity chemical grout to give enhanced shear strength, resistant to uplift. Mixing of the grouts was done at the surface and they were injected by means of proportioning pumps in the tunnels which allowed gelling times and pressures to be controlled. In addition to the main zone there were other areas treated with appropriate grouts. The main tunnel was successfully driven; air pressures were a little higher than desired, but air consumption was satisfactorily low and there were no serious losses of ground.

Ground treatment by grout injection is necessarily a costly operation in which the cost per unit volume of treated ground will vary so widely, even for use of the same grout, as to make comparisons very difficult. Material costs for acrylamide and resin grouts may be 10 times as much as for cement grouts, and require more specialised plant and highly skilled labour and supervision. The case for using a particular method needs careful study, but use of an inadequate grout that fails to effect the requisite reduction in permeability and increase in cohesion is certainly wasteful and may be positively harmful in disrupting the ground.

5.3 FREEZING

Successful freezing of permeable water-bearing ground effects simultaneously a seal against water and substantial strengthening of incoherent ground. No extraneous materials need be injected and, apart from the contingency of frost heave, the ground normally reverts to its original state. It is applicable to a wide range of soils. It takes a considerable time to establish a substantial ice wall and the freeze must be maintained by continued refrigeration as long as required.

5.3.1 History of freezing

Uses for shaft sinking through water-bearing strata are recorded in the second half of the 19th century, one of the earliest examples being a mine shaft sunk using cooled brine in South Wales in 1862. In Germany from 1883 onwards the method was developed also for mine shafts. Freezing was used in repair of a shaft on the St. Omer canal in 1893. Tunnelling and other civil engineering applications were not extensive up to 1940. It was applied to tunnelling for the Paris metro in 1907; and over the period 1932–40 in connection with ventilation shafts for the Antwerp vehicle tunnel, escalators and passages for the Moscow metro, shafts for cooling water at Swansea, and for a water supply shaft in West Cheshire.

Uses in connection with tunnelling have since then increased, as refrigeration methods and plant have been improved, and freezing is widely used in many countries.

Freezing has been employed in the construction of difficult elements of metro systems in London, Leningrad, Tokyo and Helsinki, and for sewer tunnelling in New York City and for railway tunnels in Italy. The first application of liquid nitrogen as the freezing agent seems to have been for a sewer tunnel in Paris in 1961.

5.3.2 Principles of freezing

The effectiveness of freezing depends, of course, on the presence of water to create ice, cementing the particles and increasing the strength of the ground to the equivalent of soft or medium rock. If the ground is saturated or nearly so it will also be rendered impermeable. If the moisture does not fill the pores and an ice wall is nevertheless required, either because water is temporarily absent or because it is desired to cement with ice dry running sand locally, it may be necessary to add water. The strength achieved depends on freeze temperature, moisture content and the nature of the soil. Freezing can be particularly effective in stabilising silts which are too fine for injection of any ordinary grouts.

On freezing, water expands in volume by about 9%, which does not in itself impose any serious stresses and strains on the soil unless the water is confined within a restricted volume. With water content up to about 30% the direct expansion may be up to about 3%. Frost heave, which may occur in fine silts and clays, is a slightly different phenomenon described in more detail below. In rock or clay ice lenses may build up and enlarge fine fissures so causing increase in permeability after thaw.

If there is a flow of water through the ground to be frozen the freezing time required will be increased by reason of the continuing supply of heat energy and, if the flow is large and the water temperature high, freezing may be completely inhibited. Flowing water is also liable to result in an eccentric oval rather than a circular section round the freezing tube.

As in all ground treatment techniques, adequate site investigation is necessary to allow the best system to be chosen and to design the appropriate array of freezing tubes and select plant of adequate power.

Because freezing can be imposed uniformly on a wide range of soil types in a single operation, it may offer greater security in mixed ground than treatment by injection of various grouts.

The versatility of the principle in its application to tunnel construction problems is demonstrated by the variety of its applications, such as:

1. Shaft sinking through water-bearing strata.
2. Shaft construction totally within non-cohesive saturated ground.
3. Tunnelling through a full face of granular soil.

4. Tunnelling through mixed ground.
5. Emergency local soil stabilisation.
6. Temporary underpinning of adjacent structures and support during permanent underpinning.

The cost of ground stabilisation by freezing has been generally considered to be more than for other methods of ground treatment, but this is not necessarily so and comparative estimates are often worth while. Costs of chemical injection may be difficult to estimate with any precision, because the 'take' of the ground may be unpredictable and also because small pockets of soil may require special treatment. Freezing costs in terms of plant hire and time can be estimated with considerable reliability, provided that flowing water is not present and operations are not affected by unforeseen lenses of silt or pockets of gravel.

5.3.3 Processes
Freezing may be:

1. Indirect, by circulation of a secondary coolant through tubes driven into the ground.
2. Direct, by circulation of the primary refrigerant fluid through the ground tubes.
3. Even more direct by injection into the ground of a coolant, such as liquid nitrogen.

5.3.3.1 Indirect cooling
Primary refrigeration plant is used to abstract heat from a secondary coolant circulating through pipes driven into the ground being frozen. The primary refrigerant most commonly used is Freon ($CHClF_2$) which is non-toxic, non-flammable and non-corrosive. Its boiling point at N.T.P. is -40.8°C. Other primary refrigerants are ammonia, NH_4 (-33.3°C) and carbon dioxide, CO_2 (-78.5°C), not now commonly used. The secondary coolant, circulated through the network of tubes in the ground is most usually a solution of calcium chloride. With a concentration of 30% such a brine has a freezing point well below that of Freon.

The primary refrigeration process is basically the Carnot cycle of compression and expansion reversed. The heat developed by compression of the primary coolant must be extracted by air cooling or water circulation. When water cooling is practicable the plant is relatively quiet and can be run continuously in urban areas. Efficiency is necessarily governed by the ambient temperature.

The time required to freeze the ground will obviously depend on the capacity of the freezing plant in relation to the volume of ground to be frozen and also on the spacing and size of the freezing tubes and the water content of the ground.

With a tube spacing of 1 m, cylinders of 1 m diameter or more have to be frozen, for which a typical time is 4–8 weeks. Small zones, of less than 500 m^3, can be frozen more quickly, and large volumes, over 20,000 m^3 may take up to 3 months.

5.3.3.2 Direct cooling

In these systems the primary refrigerant is circulated through the system of tubes in the ground, extracting directly the latent heat of boiling, and therefore having a higher efficiency than the indirect process.

Freon R22 is most commonly used, and it has the incidental advantage that it is non-toxic and non-flammable. It is, however, asphyxiant if it displaces air. Leakage into the ground, which is readily detected by fluid indicators or gas detectors, is normally harmless, unlike leakage of brine which makes freezing more difficult by lowering the ground feeezing temperature. Direct freezing time is similar to that for the indirect process. The choice will depend on plant availability, estimates of cost and, perhaps, personal preference rather than on any differences in ground characteristics.

5.3.3.3 Liquid nitrogen (N$_2$)

Availability of liquefied nitrogen in bulk is dependent on the existence of a regular supply to heavy industry. The major differences in this ground freezing technique are that it is not dependent on site refrigerating plant and that the temperature is very much lower and therefore quicker in application. The nitrogen under moderate pressure is brought to site in insulated containers as a liquid, which boils at -196°C at normal pressure and thereby effects the required cooling. It can be stored on site.

There is a particular advantage for emergency use, namely, quick freezing without elaborate fixed plant and equipment, and this may be doubly advantageous on sites (accessible by road) remote from power supplies. In such conditions the nitrogen can be discharged directly through tubes driven into the ground being treated, and allowed to escape to atmosphere. Precautions for adequate ventilation are important.

Where there is time for preparation, an array of freezing tubes is installed for the nitrogen circulation, including return pipes exhausting to atmosphere. The speed of ground freezing is much quicker than with the other methods, the time required being days rather than weeks, but liquid nitrogen is costly. The method is particularly appropriate for a short period of freezing, up to about 3 weeks. It may be used in conjunction with the other processes with the same array of freeze tubes and network of insulated distribution pipes, in which liquid nitrogen is first used to establish the freeze quickly and is followed by ordinary refrigeration to maintain the condition while work is executed. This can be of particular help when a natural flow of groundwater makes initial freezing difficult.

Fig. 5.5 – (a) Use of liquid nitrogen for ground freezing for a sewer tunnel.
Note: Nitrogen tank feeding array of vertical tubes (*Lilley Construction Ltd.*).
(b) Control panel regulating liquid nitrogen supply from tank behind (*A. Waddington & Sons Ltd.*).

5.3.4 Site investigation

A decision to use ground freezing methods may be taken as a result of the information obtained in the original investigation, but additional borings, samples and tests will probably be necessary to determine the best plant and arrangements. The aspects special to freezing include:

1. Extent of ground to be frozen.
2. Water content of the various materials, more particularly fine silts and clays liable to heave.
3. Salinity of groundwater as affecting freezing point.
4. Movements of groundwater.

Thermal capacity and conductivity of varying types of ground may be estimated from experience or established by laboratory tests, but the water content is the principal element, and minor variations have little effect on plant and freezing times. An ample margin of capacity is essential.

The site investigation should, of course, provide guidance in selection of the site for the plant and the distribution network, as well as the array of freezing pipes in the ground.

5.3.5 Shafts

Where a shaft is to be sunk through bad ground into firm impervious ground, an annulus only need be frozen by means of one to three rings of vertical freezing pipes installed by drilling from the surface. Excavation of the unfrozen core through the protective frozen annulus is followed quickly by building of the lining and any waterproofing, after which the ground may be allowed to thaw.

Where the shaft is entirely in water-bearing ground lacking cohesion, a plug at the bottom may need to be frozen, by means of further freezing pipes within the core. To avoid undesired and wasteful freezing of the core, these pipes can, with advantage, be insulated above the plug level. With mixed strata, pipes can be insulated at levels where freezing is not required.

Either precast or *in situ* linings can be used, and, if the freeze pattern is so designed and maintained, the whole volume can be excavated before building the shaft from the base upwards, but it is considered safer to work in units of 30 m depth or so.

Inclined shafts may be frozen either by use of inclined tubes parallel to the axis or by a carefully planned pattern of vertical tubes. Drilling of inclined holes requires specialised drilling equipment and skill to achieve the necessary accuracy.

In excavating a shaft the frozen annulus is relied on to carry the external ground and water pressures, and its thickness and temperature must be designed and ensured accordingly. The freeze must be maintained until the appropriate structural lining is completed and competent to carry the load. Any failure of the refrigeration plant allows a thaw to be initiated, with progressive danger of inundation and collapse. Where so planned and designed, however, an operation for which access is difficult in the presence of refrigeration plant but which requires only a limited time, up to perhaps 36 hours, may be executed after completion of the freeze, immediately following discontinuance of refrigeration.

5.3.6 Tunnels
The choice of drilling patterns is very similar to that for grouting. Vertical or inclined tubes so laid out as to provide a secure frozen envelope for the tunnel excavation may be driven down from the surface. Some tubes must pass through the tunnel excavation if the invert zone is to be frozen, and it is almost invariably

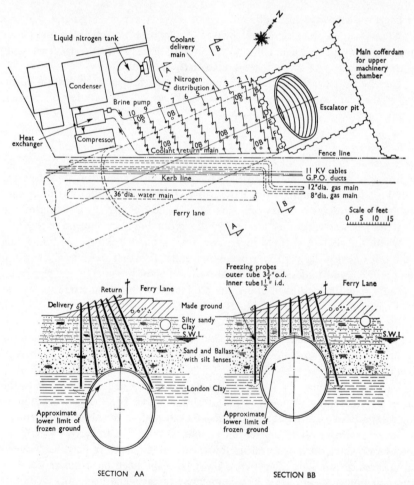

Fig. 5.6(a) – Freezing in 1966 for escalator tunnel on Victoria Line. Both liquid nitrogen, for rapid initial freezing, and ammonia-brine system were used. Site layout and sections. (*A. Waddington & Sons Ltd.*)

Fig. 5.6(b) — Freezing in 1966 for escalator tunnel on Victoria Line. (*A. Wadding-ton & Sons Ltd.*)

necessary to withdraw or abandon these immediately before excavation, which has therefore to be completed and lined without delay. The face to be excavated will be frozen unless elaborate precautions are taken, but this is usually accept-able even though blasting may be necessary. Where treatment below the invert is unnecessary, work is much simplified. Difficulties of access to the surface area above the tunnel are likely to limit severely the application of the method in cities and under rivers.

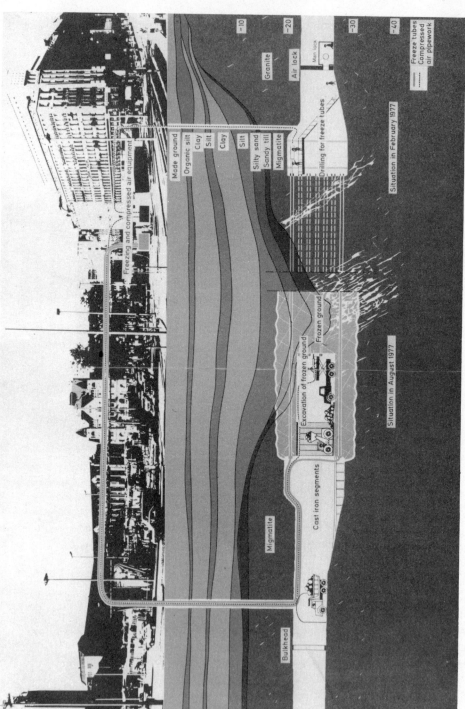

Fig. 5.7(a) — Use of freezing for tunnelling through Kluuvi cleft in Helsinki, where a gorge in the rock is infilled with waterbearing glacial deposits. Rock tunnel enlarged to permit drilling of cylindrical array of freezing tubes, preparatory to tunnelling with segmental cast iron lining. Provision was also made for compressed air working if necessary, but not in fact required. Longitudinal section.

Freezing from the tunnel ahead of the face is possible either by a conical array of pipes from the lined length of tunnel or by a cylindrical array from an enlargement or from a working pit or shaft, as for example a manhole shaft for a sewer. The former method can only be applied to a very limited length because of the spread of the pipes and the inaccuracy of horizontal drilling. it results in a very congested working area near the tunnel face, and the necessary freezing time makes the alternating progress of freezing and driving extremely slow.

The latter method of working with parallel tubes from a larger pit has substantial advantages because a longer length can be drilled and frozen and there is more working space, although this also is a slow method. Beyond about 30 m deviations of tubes are likely to increase rapidly, but it is now practicable to work up to 50 m, or in special circumstances 100 m.

Work from a tunnel face is therefore an expedient for special situations. As in shafts, the consequences of failing to maintain the freeze until the permanent lining is complete can be very serious.

Special difficulty may arise in freezing the crown of a tunnel in a river crossing where cover is shallow, because of the continuing supply of heat from the flowing water. A mat of freeze tubing installed on the river bed and covered with an insulating blanket has been used, the edges of the area being carefully sealed against hydraulic erosion.

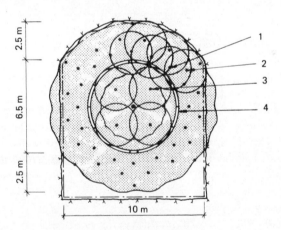

CROSS SECTION IN ENLARGEMENT
SHOWING PATTERN OF PIPES

1   INNER RING OF FREEZING TUBES
2   OUTER   ″   ″   ″
3   CORE   ″   ″   ″
4   CAST IRON TUNNEL LINING 6.1 m INT. DIA.

Fig. 5.7(b) – Kluuvi Cleft, Helsinki. Cross section in enlargement showing pattern of pipes.

The construction of tunnel junctions may be assisted by freezing the surrounding ground by tubes from the converging tunnels.

In an emergency when unexpected water is encountered locally, the merits of liquefied nitrogen freezing will be apparent.

5.3.7 Construction procedure

5.3.7.1 Tube installation

Freeze pipes consist of two concentric tubes, the outer being closed at the bottom. The coolant is circulated by pumping it down the inner tube to return through the surrounding annulus, where it extracts heat from the surrounding ground.

Spacing of about 1 m between tubes is usually appropriate, but may be made closer to provide a more uniform thickness of ice wall, or in critical areas.

Tube diameter will be chosen to suit the type of ground, length and thermal transfer requirement. Freezing will be quicker with a larger surface area for heat transfer and greater cross-section for flow of coolant. For the outer tube, the diameter is usually 50–100 mm.

Where practicable, uncased holes will be bored for subsequent insertion of the freeze tubes, but in unstable ground casings may be required, and these may be utilised as the outer coolant tube after the end has been sealed. The insertion of an independent freeze tube assembly within the casing may be preferred, especially if some lengths are to be insulated, although there will be some increased resistance to heat flow, not substantial if the gap is small. The possible leakage of coolant outwards into the ground, or of groundwater entering and contaminating the coolant must be considered. In deep shafts freeze tubes may be lubricated externally with special greases to minimise stresses imposed by heave of the ground.

Horizontal tubes from a tunnel or a pit will usually require casing and may have to be drilled through glands to prevent inflow of water. For emergency use or for short lengths, pointed tubes driven in by percussion may sometimes be practicable.

5.3.7.2 Accuracy

Spacing between tubes is determined by the need to close the ice wall as a continuous formation without 'windows'. A cylinder of frozen ground is built up round each tube as heat is extracted and a diameter of 1 m is a normal basis, which leaves some margin for deviations in the positions of adjacent tubes.

Deviation of long holes is difficult to control but it is important to ascertain the actual spacing at the ends of the borings and to ensure that tubes do not pass through the excavation zone unnecessarily.

It is therefore sound practice to survey completed holes accurately before refrigeration is started so that unsuitable holes may be abandoned, and any necessary supplementary holes may be drilled. An accuracy of about 1 in 100 is required, for which three types of instrument are used:

1. Pendulum device for vertical holes.
2. Mechanical deflection type.
3. Optical type, which employs a light beam or a laser.

These surveys are frequently carried out by specialists, using patented equipment.

5.3.7.3 Monitoring

Monitoring of the progress of freezing and its maintenance is effected by temperature readings at selected positions. The proposed outer boundary of the ice wall is likely to be chosen for points of measurement. Additional cased holes are drilled where required to accommodate thermocouples or thermistors, whereby temperature changes are measured and recorded as required. A typical provision might be five or more tubes round a shaft or tunnel, with thermistors at 3 m intervals down each tube. Observations show the progress of freezing, and circulation in particular tubes can be increased or reduced as found appropriate. Records should be maintained until the thaw is complete.

In an enclosed area, such as a shaft sealed at the bottom into clay, the water level and pressure will rise when the circumference is sealed, and this will indicate the effectiveness of the ice wall.

The refrigeration plant should be equipped with thermometers, pressure gauges and flow meters to evaluate heat extraction, and with integrating meters for power consumption to allow a check on overall performance.

It may be advisable to coordinate temperature records with levelling at the surface, to monitor any ground movements resulting.

5.3.7.4 Excavation

Soft ground frozen to a temperature of $-10°C$ to $-15°C$ has strength similar to weak concrete or soft rock, about $10-20$ N/mm^2.

For hand excavation the heavier types of pneumatic tool are required, fitted with points. Soft ground tunnelling machines may be inadequate, and other mechanical excavators must be matched to requirements.

Excavation by drilling and blasting is often considered best, with charges carefully designed to avoid damage to tubes and minimise vibrations, with due regard to the limited zone of ice.

Excavation can be extended in advance of lining if a sufficiently thick zone is maintained fully frozen, but frozen ground can creep under load if support by lining is too long delayed.

5.3.8 Frost heave

Expansion and frost heave have already been mentioned. The immediate volumetric expansion of 9% which occurs as water solidifies is a limited action only relevant in an enclosed volume from which the liquid water cannot flow out freely. Frost heave proper is typically a surface phenomenon occurring naturally in cold climates where an actively freezing boundary in fine grained water-bearing soil can draw to itself a continuing supply of water, by capillary flow, whereby a progressively thickening layer of ice builds up by crystalline growth and forces up the overlying soil. In artificial freezing for shafts and tunnels it may likewise occur, but the supply of water need not be restricted to capillary flow. Enlargement of fissures may result, contributing to heave.

These phenomena are particularly related to silts whose analyses show a few per cent to have grain size less than .02 mm. From coarser soils water can be expelled more freely and finer clays become impermeable to the movement of moisture. Chalk can show heave, as can fissured clays.

Water pressure build-up may be observed if a zone being frozen is enclosed within an impervious envelope, which may be the ice envelope itself, or may be clay. Bursting action can be prevented by provision of pressure relief pipes, provided they are not allowed to freeze solid prematurely. The development of water pressure within the zone is a positive indication of the continuity of the ice envelope.

On termination of freezing the actions of expansion and heave are reversed as the ground thaws, which may occur naturally over a period of 6–20 weeks, depending on volume frozen, groundwater flow, ambient temperature. Thaw can be accelerated by circulation through the freeze tubes of steam, heated water, or brine at normal temperature. It has been noted in some cases where heave has been recorded that subsequent settlement has been greater by up to 20%.

Services such as water mains which might be damaged by freezing and heave can be protected by excavation to below their level.

5.4 DEWATERING

The management of water during excavation has already been discussed in Volume 1, but the problems arising at that stage can be minimised by pre-liminary dewatering of the ground, so as to improve the stability of the tunnel face and provide drier and safer working conditions.

In permeable strata where k exceeds about 10^{-3} cm/s, or where an aquifer can be dewatered below less permeable strata, the level of the water table over a wide area can be drawn down by pumping from boreholes and deep wells. These processes are widely used in open excavations and are suited also to cut-and-cover tunnels and shallow bored tunnels, but less so to deep bored tunnels.

The two principal methods are well points and deep filter wells. There are also tunnel drainage techniques of less general application such as pilot tunnels or headings excavated below and in advance of the main tunnel, and sub-drains and sumps maintained below the invert as the work advances.

In special circumstances in very fine grained soils vacuum consolidation and electro-osmosis might be considered.

5.4.1 Well points

Where an area of permeable ground is to be dewatered to a depth of less than about 6 m—limited by the effective vacuum lift of a pump—a ring of tubular well points of about 125 mm diameter can be sunk by wash boring round the perimeter of the area at intervals of a metre or more. The lowest part of each tube is perforated and fitted with a gauze strainer; at the surface the tubes are connected through a ring main to a suction pump. Some days of pumping may be necessary to draw down the water within and immediately outside the ring, followed by continued pumping to maintain the low level by intercepting inflowing water. Greater depths can be attained in conjunction with excavation in stages of about 5 m by installation of further inner rings of well points at the successive lower levels, while maintaining pumping in the original ring.

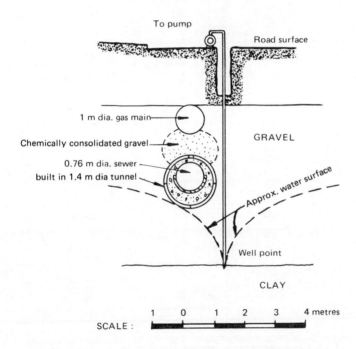

Fig. 5.8 – Example of well points used in combination with grout injection: Extension of London Central Line 1939. Diversion of 0.76 m dia. sewer was effected in a 1.4 m dia. C.I. lined tunnel, driven in free air through water bearing gravel, beneath an obliquely lying 1 m dia. gas main. Chemical grouting (to secure the ground and support the gas main) was injected progressively ahead of the tunnel working face.

5.4.2 Deep wells

As the name suggests, this method is for work at greater depths. A number of larger diameter wells, 300 mm or more, are sunk at spacings of 3 m or more in a ring to below the level required. Properly graded gravel filters are formed at the bottom of each bore around perforated suction pipes, and submersible pumps are installed at the appropriate level within the casings. In principle the operation is similar to that for well points, but the wells are larger, fewer and more widely spaced and are not limited in lift by considerations of vacuum suction. Because of the wider spacing the local draw down at the well must be further below the required working level.

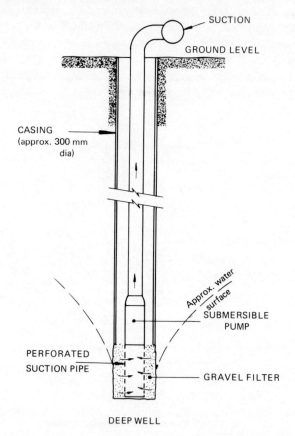

Fig. 5.9 – Deep well for ground water lowering below about 6 m.

5.4.3 Settlement

In both systems the filter arrangements are important to avoid settlement caused by drawing out fines from the soil. Even without any loss of fines there may be consolidation resulting in settlement of the ground affected. There is potential damage to foundations of buildings and structures over a wide area. Wells, springs and other sources of groundwater may be detrimentally affected, either temporarily or permanently. Near the sea, saline water may be drawn in. In such cases it may be possible to avoid the effects by recharging, through feed wells, important zones outside the area of excavation simultaneously with the operation of pumping down.

5.4.4 Pilot tunnels
Drainage pilot tunnels are an old and familiar expedient particularly relevant to subaqueous tunnelling both to improve conditions for excavation in the main tunnel and possibly to provide a permanent drainage system. The Seikan tunnel (Volume 1, Chapter 10) is an outstanding modern example of this method.

5.4.5 Vacuum consolidation
The method is applied in fine grained soils, such as silts, of low permeability. It is an extension of the well point system, in which a vacuum is created in the well points and the surrounding silt which is consolidated by the differential effect of atmospheric pressure at the ground surface.

5.4.6 Electro-osmosis
An electric current between electrodes driven into saturated fine grained soil will cause pore water to move from anode to cathode. If the cathode is constructed as a well point, seepage water can be pumped away progressively and consolidation of the soil will result.

BIBLIOGRAPHY

The subject is one continuously growing in its application to all forms of civil engineering construction. In soft ground applications it is almost a specialised branch of soil mechanics. The special aspects relevant to tunnelling are mostly discussed in papers descriptive of individual projects.

Mussche, H. E. and Waddington, J. C., Applications of the freezing process to civil engineering works, *Instn Civ. Engrs Wks Constr. Div.*, 1946, (5).

Kell, J., Pre-treatment of gravel for compressed air tunnelling under the River Thames at Dartford, *Chart. Civ. Engr.*, 1957 (Mar.).

Glossop, R., The invention and development of injection processes, *Geotechnique*, 1960, **10** (Sept.) and 1961, **11** (Dec.).

Ischy, E. and Glossop, R., An introduction to alluvial grouting, *Proc. Instn Civ. Engrs*, 1962, **21** (Mar.); *see also* discussion, 1962, **23** (Dec.).

Mansur, C. I. and Kaufman, R. I., Dewatering, Ch. 3 in Leonards, G. A. (Ed.), *Foundation Engineering*, McGraw-Hill, 1962.

Kell, J., The Dartford Tunnel, *Proc. Instn Civ. Engrs*, 1963, **24** (Mar.); *see also* discussion, 1963, **26** (Dec.).

Grouts and drilling muds in engineering practice, Butterworths, 1963.

Perrott, W. E., British practice for grouting granular soils, *Proc. Amer Soc. Civ. Engrs, J. Soil Mech Fndtn Div.*, 1965, **91** (Nov.).

Kell, J. and Ridley, G., Blackwall tunnel duplication, *Proc. Instn Civ. Engrs*, 1966, **35** (Oct.); *see also* discussion, 1967, **37** (July).

Dempsey, J. A. and Moller, K., Grouting in ground engineering, *Proceedings Ground Engineering Conference*, London, 1970.

Cashman, P. M. and Haws, E. T., Control of groundwater by water lowering, *ibid.*

Sanger, F. J. and Golder, H. W., Ground freezing in construction, *Proc. Amer. Soc. Civ. Engrs, J. Soil Mech Fndtn Div.*, 1968, **94** (Jan.).

Collins, S. P. and Deacon, W. G., Shaft sinking by ground freezing: Ely Ouse — Essex scheme, *Proc. Instn Civ. Engrs*, 1972, **Suppl.**, p. 129; *see also* discussion, p. 319.

Powers, J. P., Groundwater control in tunnel construction, *Proc. 1st Rapid Excavation and Tunneling Conf.*, Chicago, 1972, **1**, p. 331.

Shuster, J. A., Controlled freezing for temporary ground support, *ibid*, **2**, p. 863.

Bowen, R., *Grouting in engineering practice*, Applied Science, 1975. 2nd edn., 1981.

Bell, F. C. (Ed.), *Methods of treatment of unstable ground*, Butterworths, 1975.

Clough, G. W., *A Report on the practice of chemical stabilization around soft ground tunnels in England and Europe*, Dept. of Civ. Engng, Stanford University, 1975.

Herndon, J. and Lenahan, J., *Grouting in Soils*, 2 vols., U.S. Federal Highway Administration Reports FHWA-RD-76-26 and 27 (NTIS PB 259043 and 4), 1976.

Jones, J. S. and Brown, R. E., Temporary tunnel support by artificial ground freezing, *Proc. Amer. Soc. Civ. Engrs, J. Geotech. Engng Div.*, 1978, **104** (Oct.).

International Symposium on Ground Freezing, Bochum, W. Germany, 1978.

Jones, J. S. and Brown, R. E., New advancements for ground freezing for tunnel construction, *Proc. 4th Rapid Excavation and Tunneling Conf.*, Atlanta, 1979, **1** p. 722.

2nd International Symposium on Ground Freezing, Trondheim, Norway, 1980.

Powers, J. P., *Construction Dewatering, a guide to theory and practice*, J. Wiley, 1981.

6

Highway Tunnels

6.1 HIGHWAY REQUIREMENTS

The special construction features for highway tunnels include geometrical configuration, road construction, lighting, ventilation, traffic control, fire precautions, general facilities for cleaning and maintenance, and often provision for the pipe and cable services of public utilities. The requirements will vary according to the tunnel situation and character; whether urban, motorway, subaqueous or mountain, and whether long or short.

A highway tunnel is part only of a larger highway system. Its function in providing a crossing under an obstacle may be of major importance in linking up areas with poor communications, as under an estuary or through a mountain range. In some instances, tunnels may be relatively minor elements as in a road following a rocky gorge.

Road systems are designed on the basis of their traffic capacity, taking account of lane widths and clearances, gradients and traffic composition. The same formulae can be applied to tunnels, but it has been found that in tunnels where regular commuters constitute a substantial proportion of peak hour traffic the peak hourly flow may exceed the nominal capacity by 50% or more.

For a number of two-lane one-way tunnels actual flows of from 3000–4500 vehicles/hour have been observed as compared with nominal capacity from 2000–2700. In contrast, rural and mountain tunnels, in whose peak traffic the proportion of regular users is small, may have an actual capacity of 80–90% of the nominal capacity.

The principal factor determining the traffic capacity of a tunnel is the geometry of its alignment and cross section, but its lighting and ventilation may impose lower limits. Special factors influencing traffic flow include freedom from side turnings, absence of cyclists and pedestrians, and the regularity of use by a substantial proportion of drivers, or, alternatively, unfamiliarity. If capacity is found less than adequate the geometry cannot be readily altered, but deficient lighting can usually be improved to any necessary extent. Ventilation may be boosted, but major improvements are likely to be costly in construction and operation.

In all cases the geometry of the tunnel alignment is determined by the siting of the portals, and by the curves and gradients adopted to connect them. As part of a highway system the alignment must conform to the line of the highway as a whole, and must be compatible with the standards adopted. A tunnel, other than a very short one, is likely to be a substantial element in the functioning of the highway and in the construction cost. The highway standards must then be formulated in the context of the practical possibilities of tunnel construction. A short motorway tunnel through a ridge will, as nearly as practicable, be laid out to the ordinary motorway standards, but a subaqueous crossing under an estuary, or a long mountain tunnel will be sited primarily to suit tunnelling conditions and will largely determine the layout and standards of the approach roads connecting it to other parts of the highway system.

6.1.1 Urban underpasses

In urban conditions, for relatively short underpass tunnels, the points of connection into the road system are likely to be very narrowly limited, and further limitations on the tunnel line will be imposed by existing buildings and by buried services and structures. For underpasses following existing alignments, wide roadways are needed in order to provide space for separate descending approaches to the portals. The tunnels will usually be shallow and may often be constructed by cut-and-cover methods, but occasionally, as when at greater depth under a hill, by boring. The straightest practicable line will be adopted and gradients will be restricted if possible to less than 3%, because steeper gradients give rise to congestion when large heavily loaded vehicles are ascending.

The concept of a network of major traffic routes deep beneath a city has been advanced but presents exceptionally formidable problems of cost of construction, ventilation, intermediate access, fire and flood, and will not be further discussed here.

6.1.2 Motorways

Motorway tunnels are relatively new. In open country the obstacle requiring tunnelling may be a ridge, or hill spur, or occasionally the crossing of another road or rail route where viaduct construction is impracticable, or even the crossing of park land where the highway has to be sunk for reasons of amenity. In all these the tunnel line and gradients must conform with the standards specified for the motorway: sight lines appropriate to the design speed may require particular care, especially where vertical curves are necessary. Adequate lighting may be a particular difficulty. One possible relaxation for tunnel construction may be the omission of an emergency lane as provided on the 'hard shoulder', but in long tunnels lay-bys for use in emergencies might be a valuable contribution to safety and convenience. Limitations on speed may have to be considered.

Fig. 6.1 – Layby at mid tunnel. Mersey Kingsway tunnel was provided with an extra lane width over a length of 30 m, by construction of a special elliptical tunnel section.

6.1.3 Subaqueous tunnels

For subaqueous tunnels the line will usually be fixed nearly at right angles to the waterway to minimise the length of tunnel, unless the topography of the valley imposes another alignment. Roads tend to follow the line of the valley and right angled junctions are likely to be required between the existing roads and the tunnel approaches.

Steeply rising river banks where the climbing gradient out of the tunnel does not adequately exceed the topographic gradient make the approaches difficult. As an example, the Tyne tunnel (1967) on the north bank of the river gains the necessary height by making a turn of 200°, at the exceptionally small radius of 128 m, in rising to the portal at a gradient of 5%, and the emergent approach road completes the loop by crossing above the tunnel. As with urban tunnels any gradient exceeding 3–4% slows heavy traffic disproportionately. The Clyde tunnel, which is an urban subaqueous tunnel, has had to adopt gradients of 6% in order to connect into existing highways, close to and parallel to the river and therefore at right angles to the tunnel.

The profile of a subaqueous tunnel will usually comprise a descending gradient, a nearly level central length and a rising gradient. Vertical curves, required at the changes of gradient, should be as long as practicable, both to simplify construction if a shield is used and to avoid restricting unduly the sight line for vehicles in the tunnel approaching the change of gradient.

The depth of a subaqueous tunnel is determined by the bed level of the waterway and the geology of the river bed, together with the cross section of the tunnel. A circular cross section is usual for a bored tunnel, and this requires excavation at the crown substantially above the vehicle clearance level. In addition a safe thickness of undisturbed ground above is required during construction, depending on the character of the rock and soil and the construction methods adopted. In the case of navigable waterways, future dredging plans need to be considered. The first Blackwall tunnel (1897) was successfully driven with a specially designed shield and compressed air, with a mere 1.6 m of gravel and silt, but supplemented by 3 m of clay blanket, above the hood at mid-river (see Volume 1, p. 36). For the second Blackwall tunnel (1967) cover was about 6 m and the ground was extensively grouted. Again a shield and compressed air were used.

6.1.4 Mountain tunnels
Major mountain crossings also impose their own rules. A pass, or col, approached by a steep and narrow valley can only be crossed by a surface roadway by use of very steep gradients, sharp bends, and zig-zags. Such passes are also liable to be closed or made dangerous by snow and ice and possibly by rock falls and avalanches. The shortest practicable tunnel under the col may ameliorate the worst conditions, but a longer tunnel at a lower level makes possible a much better highway.

The cost of any such tunnel is so high that a substantial traffic expectation is needed to justify the expenditure, but if sufficient demand exists it is likely to be stimulated by improved facilities, so that a major project including improved approach roads and a long tunnel at a lower level may well be found advantageous. The whole layout of the highway system for such a crossing involves much study of topography and geology, and balanced judgement in the choice of alignment and design of viaducts and minor tunnels in the approach roads as well as in the major tunnel.

Although there is more freedom of alignment for a road than for a railway, because steeper gradients and sharper curves are acceptable, the problems remain difficult and complex. To the limitations of longitudinal gradient and curvature must be added in greater degree the transverse slope of the ground, which is of increased importance where wide carriageways are required. A narrow road can be accommodated on sidelong ground with a limited need for cutting and embankment, whereas a major highway with dual two- or three-lane carriageways cannot. With any specified gradient it is geometrically possible even on the steepest slopes to set out a centre line ascending by sharp zig-zags, but width of construction and minimum radius for curves make such routes quite impracticable if the slope is excessive. Tunnels through projecting spurs and under the summit col become necessary.

The comprehensive design of highway crossings for mountain routes is too large a subject to be dealt with in adequate detail here. Their geometry must be related to the topography and the geology in order to design, and ensure the stability of cuttings, embankments, viaducts and portals leading to tunnels.

Fig. 6.2 – Canopied entrance to Mont Blanc Tunnel – West portal, France.

6.2 GEOMETRY OF CENTRE LINE

The geometry of any tunnel has two aspects, that of the centre line and that of the cross section. They are not entirely independent but can best be considered separately.

The principal factors determining the centre line include:

1. The relative positions of the portals and directions of approach
2. Geology
3. Clearances from external obstacles
4. Gradients
5. Vertical curves
6. Horizontal curves

6.2.1 Approaches

The obvious initial approximation of aligning the tunnel in a straight line joining the portals will almost inevitably be modified in any but a very short and simple

tunnel by introduction of curves to suit the approaches, and varying gradients
to carry it under and around obstacles.

6.2.2 Geology
The choice of the most suitable strata for tunnelling may influence the alignment
and level greatly, as may the avoidance of water-bearing ground or unstable rock.
The subject is dealt with in Chapter 4.

6.2.3 Clearance from external obstacles
Reference has already been made to the clearance required below the river bed
in a subaqueous tunnel. Adequate clearances from existing structures are also
necessary in urban conditions. As a broad generalisation it is usually satisfactory
if uniform undisturbed ground outside the tunnel extends for one tunnel
diameter; if discontinuities or obstructions occur within this zone, more careful
analysis and special precautions are likely to be required.

6.2.4 Gradients
Although road vehicles are normally capable, even when fully loaded, of climbing
steep gradients, this is not an acceptable basis of design for highway tunnels.
Heavy vehicles on steep gradients are reduced to the use of their lowest gears,
and can only proceed at a road speed of perhaps 10 km/hr, which is extremely
restrictive of other traffic and greatly reduces the traffic capacity of the tunnel
and its efficient use, at the same time making abnormal demands on the
ventilation system. While limitation of gradients to 3% is desirable, up to 5%
of even 6% is acceptable, particularly if two traffic lanes in the direction of
climb are available. In short urban underpasses steep gradients may be necessary
where space is very restricted. If their function is relief of rush hour congestion
and their traffic is mainly light vehicles, little delay may in fact be experienced.
In long mountain tunnels it will usually be possible so to select portal positions
that the gradients are within the desired limits. On the other hand, in subaqueous
tunnels the length and cost of the scheme will be minimised if the maximum
practicable gradient is accepted. One expedient adopted on the Mersey Kingsway
and Dartford II tunnels has been the combination of a two-lane tunnel with a
three-lane open approach on the ascent, to provide a 'crawler' lane for slow
and heavy vehicles, the gradients being 4% both at Mersey and at Dartford.

A minimum gradient should also be specified, to ensure longitudinal drainage
of the roadway. For most conditions 0.25% is probably acceptable.

6.2.5 Vertical curves
Changes of gradient are normally small in motorway and mountain tunnels and
connecting curves are correspondingly short and can follow the ordinary highway
specification.

In subaqueous tunnels and in urban underpasses, however, gradients of 5% or thereabouts are normal and the junction between this slope and the central level length should be eased by a curve, whose radius will be determined partly by the length available and partly by the method of construction. In a shield driven tunnel with precast segmental lining the steering of the shield must be gradual, and the segmental rings must either be designed as tapered rings to suit a specified radius or adapted to the curvature by insertion of tapered packings in the joints. In a tunnel of 9 m diameter using 1 m rings, a radius of 1000 m requires a taper of 9 mm per ring. If specially cast taper rings are employed they will probably have a taper of about 50 mm and therefore only one special ring in five or six need be used.

Such a curve is much sharper than is usual in highway practice, and would not be acceptable in a motorway tunnel, but may be appropriate in a subaqueous tunnel in which speed is limited, and may be essential in an underpass where length is very limited. It may be noted that, on the open highway, minimum radii for convex vertical curves are determined by sight lines appropriate to design speed, and for concave curves by vertical acceleration. In a tunnel the concave curve will also restrict the sight line because a low ceiling line will limit visibility ahead.

6.2.6 Horizontal curves
In plan, curves may be necessary to align the tunnel with its approach roads and to avoid obstacles in the ground. The same considerations apply in determining the radius as in surface roads: design speed, centrifugal force, superelevation, line of sight.

On very sharp curves, some extra lane width for long vehicles is desirable, but may be impracticable or prohibitively expensive. A special consideration, particularly where the traffic emerges on a rising gradient—as is normally the case in a subaqueous tunnel—may be the avoidance of an alignment directly towards the sun when it is low in the sky. A curve in the open cut approach, or a screen wall, may be found helpful in minimising dazzle.

6.3 CROSS SECTION

The cross section is determined by the space required for traffic, and other facilities, and by construction methods.

6.3.1 Traffic space
The traffic space required is defined by the width of lane and maximum load height of vehicle. The minimum normal tunnel will accommodate two lanes of traffic. Three lane tunnels are not uncommon where a rectangular section is used, in cut-and-cover construction or in immersed tubes. The Mersey Queensway tunnel (1934) was unusual in providing for four lanes of traffic

in a circular bore, but the lane width is only 2.74 m and the side lanes have a height limitation at the kerb of 4.42 m. The more common pattern for bored tunnels is to construct them in two-lane widths, as in the second Mersey Kingsway tunnel with two tubes, or the Elbtunnel with three tubes.

For vehicular traffic the equivalent of the 'structure gauge' on the railways is a rectangle, defined by lane widths and height, plus such clearances as are appropriate for road vehicles, which are not under the same precise control as rail vehicles, in construction, loading, steering and suspension.

There has been a progressive increase in traffic lane widths and heights, and consequently in tunnel diameters since the end of last century which can be seen in the following table of dimensions of British subaqueous tunnels of circular section. In such tunnels the dimensions are more closely restricted to the minimum acceptable than in mountain tunnels or cut-and-cover construction because any increase adds disproportionately to cost and difficulty. In some cases, roadway dimensions have been determined by commitments entered into many years before actual construction. Thus the Dartford tunnel was planned and shields were built and installed in 1938 to the standards then current, but work on the main tunnel did not proceed until 1957 and 6.40 m was the widest roadway that could be fitted into the 8.59 m circle, even with slightly substandard headroom of 4.88 m. The same shields and cast-iron lining were then used at the Blackwall II tunnel, the width being reduced to 6.10 m, but with full headroom of 5.03 m.

BRITISH SUBAQUEOUS HIGHWAY TUNNELS								
Tunnel	Construction	Internal diameter	Length between portals	Lane widths	Height at Kerb	Max.	Max. gradient %	
Blackwall I	1892–97	7.52	1361	2x2.44			2.9	
	1967–68			2x2.97	3.23 4.35	4.41		Improvements
Rotherhithe	1904–08	8.43	1480	2x2.44	4.88	5.64	2.7	
Mersey Queensway	1925–34	13.41	3237	4x2.74	4.42	7.16	5	
Dartford I	1937–38							Pilot only
	1957–63	8.59	1429	2x3.20	4.88	5.33	3.6	
Clyde	1961–64	8.99	747	2x2x3.35	4.57	5.03	6.3	Twin tunnels
Tyne	1961–67	9.53	1684	2x3.65	4.88	5.26	5	
Blackwall II	1960–67	8.59	1174	2x3.05	5.03	-	4	Duplication Blackwall I
Mersey Kingsway	1966–71	9.63	2256	2x2x3.65	5.03	5.49	4	Twin tunnels
Dartford II	1973–80	9.54	1435	2x3.65	5.03	-	4	Duplication Dartford I

The current British standard clearance for height is 5.03 m, which is greater than the ordinary minimum of 4.5 m on the European continent, or 4.3 m in North America.

With the circular form the rectangular traffic area occupies no more than about 60% of the area of circle for a two-lane roadway, but footways for inspection and maintenance are usually accommodated at the sides, while the space beneath the roadway is very valuable as a ventilation duct and for various services, and space above any ceiling can also be used for ventilation and cables. Although surprisingly economical for two lanes, the circular form does not lend itself so well to three or more lanes, the traffic area falling below 50% of the circular space.

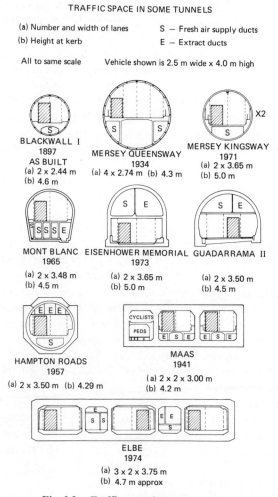

Fig. 6.3 – Traffic space in some tunnels.

6.3.2 Other space

Walkways are often considered essential for inspection and maintenance, and for emergency use both for access to the site of an accident and for escape. They have to be protected from impact by vehicles, and can with advantage be elevated above road level and also protected by a kerb and a safety margin. Additional space may also be necessary for ventilation ducts. In a circular tunnel the spaces beneath the roadway and above the clearance line are available without extra excavation, and in a horseshoe tunnel there is normally a substantial area in the crown, but in a rectangular structure extra width or depth must be excavated. This subject is more fully discussed in Chapter 9.

Other provisions often required, particularly in river crossings, are the accommodation of water mains, gas mains, electric power cables, or other services. In a circular tunnel these can usually be installed in the under-road space but at the expense of reducing the area available for ventilation and increasing the necessary fan power to overcome the friction and turbulence generated. Gas mains may require special study of the contingency of leakage or of a burst filling the tunnel with a toxic and explosive mixture. Electric power can only be transmitted at the expense of heat losses, and these may have a significant effect on raising the air temperature within the tunnel, possibly to an unacceptable level in extreme conditions. The risks of accidental damage to any of these services and the consequences to tunnel users need to be assessed carefully.

6.3.3 Cyclists and pedestrians

In the construction of some tunnels, particularly urban subaqueous tunnels, there is a demand for crossing facilities for pedestrians and cyclists. These can be disproportionately costly if incorporated in a vehicular traffic system. One solution, adopted on the river Tyne, is the provision of a separate smaller tunnel for pedestrians and cyclists. In the Clyde tunnel a cycle track and walkway were accommodated within the main tunnel, below the road deck, but the headroom over the carriageway at the kerb line had to be made substandard, namely, 4.57 m, and the fresh air provisions for the subsidiary passage had to be to a very high standard because the intake of CO by those taking physical exercise is more rapid and the exposure is longer than for those driving through. In the Dartford tunnel a transporter vehicle was provided for cycles and no pedestrians were admitted.

Other facilities, in addition to ventilation, to be incorporated within the tunnel are the services for the tunnel itself: lighting, emergency services such as telephones and fire alarms, fire mains, meters for carbon monoxide and visibility, public address systems, traffic lights and signals, drainage and pumping.

6.3.4 Construction requirements

The shape of a highway tunnel, whether rectangular, horseshoe or circular in form, is largely dictated by the method of construction adopted to suit the ground conditions. For cut-and-cover the rectangular shape is usual, for rock to be excavated by blasting, the horseshoe or other arched form, for excavation by full face machine the circle, and likewise for most soft ground subaqueous tunnels (other than submerged tunnels). In long mountain tunnels a rising gradient is preferred to simplify drainage during construction; in shield driven tunnels sharp curves, horizontal or vertical, present considerable difficulties in steering the shield and building the lining.

6.4 ROAD CONSTRUCTION

In a tunnel through sound rock, road formation is unlikely to present any unusual problems. Drainage of ground water seepage is, of course, important and the gradients and cross falls of the road surface must be accurately followed. Specified headroom may require particular care at the kerb where the tunnel roof arches down.

In a circular tunnel the roadway is necessarily suspended above the invert and the structural slab may be of reinforced concrete, precast or *in situ*, or of composite prestressed concrete beam and *in situ* slab construction. It must be very accurately positioned in relation to the tunnel lining.

The programming of road construction in a tunnel is not easy, unless it can wait for completion of the tunnel excavation and lining, because the road works obstruct access for any other supplies to and from the working face.

Temperature movement is not normally a serious problem in a long tunnel, because fluctuations are smoothed out, even round the year, in that the tunnel lining and surrounding ground act as a heat reservoir.

Because of the very limited clearances, usually minimal above the kerb lines, the kerb should be sufficiently high to ensure that it cannot easily be mounted by the wheel of a vehicle, and should also be robust. In setting out the kerb line it may be necessary to prepare and follow a 'wriggle diagram', as described in Volume 1, to accommodate unavoidable deviations in the tunnel alignment.

For the surfacing of the roadway, accurately laid concrete in the road slab without bituminous topping may be specified, but it must be remembered that ordinary highway paving trains cannot usually be operated within the confines of a tunnel. An asphaltic or bituminous wearing coat is most common in practice. It should be to a high standard. It may be noted that it will not be exposed to extreme temperatures and to sunlight. As in the case of concrete work ordinary paving equipment cannot be used, and also ventilation during laying must be ensured. Eventual resurfacing needs consideration; the old wearing coat will almost certainly have to be stripped off because headroom must be maintained and a shut down of the tunnel is likely to become necessary.

Fig. 6.4(a) & (b) – Road construction in first Dartford Tunnel. Precast units assembled in tunnel invert.

6.5 SECONDARY LININGS

It is usual to provide a secondary lining to the side walls of a highway tunnel to contribute to visibility and for aesthetic reasons, and sometimes to assist in waterproofing. The roof may also be similarly treated, but in many cases a false ceiling is constructed to form ventilation ducts and cable runs above, or to shed water or for acoustic absorption. In all but very short tunnels, artificial lighting is considered necessary and its efficiency is greatly enhanced by suitably light coloured reflective surfaces. In most tunnels some seepage of water through the primary structural lining is inevitable and even very slow penetration of groundwater can discolour large areas, both by direct deposit of dissolved salts and by assisting the adhesion of particles of soot and oil from vehicle exhausts. An air gap, or a waterproof membrane between primary and secondary linings can be helpful in preventing this seepage staining.

The principal criteria in the selection and design of such linings for tunnel walls are:

1. Costs of installation, maintenance and replacement.
2. High light reflection, but diffuse not specular.
3. Non-absorbent surface resistant to water, oil and dirt; washable with detergents and not readily scratched.
4. Resistance to aging and corrosion in tunnel atmosphere.
5. Rigid dimensionally and free from vibration.
6. Fire resistant and not contributory of noxious fumes in fires.
7. Replaceable when damaged and removable where required for access to services behind.
8. Minimum thickness, shaped to profile.

Many different forms of internal finish have been used, although in some tunnels bare rock, or unpainted or painted concrete or cement rendering are found adequate. Finishes include white glazed tiles, coloured tiles, vitreous mosaic tiling, vitreous enamelled steel sheeting, plastic coated aluminium sheeting, PVC sheeting. Higher aesthetic standards are to be expected in a major city tunnel than in an industrial or dock environment or in a mountain area.

The merits of ceramic or vitreous tiles are apparent, but they present problems of adhesion and cracking. They are often fixed direct to a primary concrete lining without an air gap, which makes it difficult to control leakage and staining at joints. Repair of damage should not be unduly difficult.

Coated steel sheeting can be prefabricated in large panels with some form of framing to provide rigidity and to make fixing and replacement easy. Access to cables and other services can readily be given. Vitreous enamelling provides a hard washable non-absorbent surface but it can be damaged by impact. Plastic coating is of course less brittle but more easily scratched and possibly more absorbent.

Fig. 6.5 – Examples of secondary linings.

Fig. 6.5(a). Vitreous tiles on concrete wall filling to cast iron lining in first Dartford Tunnel. Acoustic lining to ceiling.

Fig. 6.5(b) — Vitreous enamelled steel sheeting to walls in Tyne Tunnel.

Fig. 6.5(c) — Vitreous enamelled steel sheeting forming cable ducts in walls of Mersey Kingsway Tunnel. No other secondary lining is employed but epoxy resin coating and paint are applied directly to inner steel facing of composite steel and concrete segments.

Plastic panels are lighter, usually cheaper and more easily adapted to fit where there are irregularities. They are also more readily scratched, possibly more absorbent, and are sometimes rejected as an unacceptable fire hazard, particularly in respect of fumes generated.

Roof or ceiling linings present similar problems. Reflection of light is not considered to be of the same importance as for walls. Indeed, black ceilings have been installed, one argument being that there is created for the driver a greater impression of overhead space. Sound reverberation in a tunnel is sometimes considered to be so objectionable as to justify the installation of acoustically absorbent linings. A false ceiling can be used for this purpose, but absorbent materials, such as slag wool, are essentially open textured and not washable. Therefore such a ceiling will inevitably become darkened by smoke which will be least objectionable on a dark surface. Much of the sound reflected directly from the walls is not attenuated by acoustic treatment of the ceiling.

In the absence of a false ceiling the crown of the tunnel may be treated in the same way as the walls, but a less costly finish is likely to be adopted, possibly cement rendering and painting of the primary lining surface. The problem of cleaning has to be considered.

It is usually necessary also to accommodate in the ceiling or at its junction with the walls, the lighting units and their feed cables, and in some cases jet booster fans are installed above the vehicle clearance level.

6.6 LIGHTING

The inadequacy of natural lighting for any but a very short tunnel is obvious, and the design of lighting on a safe and economic basis is complex. It must be suited to the performance of the human eye, subjected to varying conditions of light, and must ensure that the roadway with guide lines and any obstacles on it can be seen and appreciated at a sufficient distance ahead, appropriate to the vehicle speed and braking distance.

The subject of highway tunnel lighting is dealt with in some detail in the periodic publications of the Permanent International Association of Road Congresses (P.I.A.R.C.), the latest of which relate to the XVI Congress held in Vienna in 1979.

The units employed below in describing lighting are for the *luminance* or brightness of any illuminated surface and *illuminance* or flux of light directed onto the surface.

For the purpose of tunnel lighting they are connected by the simplified formula

$$L = \frac{E \times P}{\pi}$$

where: L is the luminance measured in candelas/m^2
E is the illuminance measured in lux
P is the coefficient of diffuse reflection of the surface,
which lies in the range 0.1 to 0.3 for various carriageway
surfaces.

6.6.1 Optical adaptation

Levels of lighting, adequate at night on the open road, whether from overhead lighting or from vehicle headlamps, are totally inadequate for a driver entering a tunnel from daylight. The comparative luminances (measured in candelas per square metre) range from 8000 cd/m^2 or more in an open sunlit approach to about 2 cd/m^2 or less for good highway lighting at night. The inadequacy arises from the fact that the eye at any time is temporarily adapted to a certain average brightness and can discriminate only over a limited range above and below this average. It suffers from glare from much brighter light and cannot discern any detail below about 5% of its adaptation level. The eye does, however, change its adaptation level fairly rapidly in response to a change in lighting. At moderate levels of brightness it will accommodate to a decrease in the luminance of the field of vision by about half in every 3 seconds. In terms of vehicles travelling at a speed of 72 km/hr this corresponds to halving the luminance every 60 m. This progressive reduction cannot be continued indefinitely even in a long tunnel, the rate of adaptation at low levels being much slower. At the exit from a tunnel adaptation to brighter light is much more rapid.

At any point in a vehicle's entry to a tunnel two levels of lighting are significant: first, the general luminance of the driver's field of vision to which his eye has adapted, and second, the luminance of the roadway at a safe distance ahead so that the vehicle may be stopped if an obstacle is seen. The discussion that follows relates primarily to long tunnels, but the same principles apply to short tunnels, except those so short (less than about 40 m) that the exit is clearly visible before entry.

6.6.2 Entrances

The origin for calculation of lighting levels at entry to a tunnel is a point in the approach road at the safe stopping distance from the entrance. This may be of the order of 100 m or more depending on speeds, gradients and road surface. From this point the eye should be able to see any obstacle within the tunnel entrance, which should not simply appear to be a black hole. At this origin maximum luminance in sunlight may be between 2000 cd/m^2 where there is a hilly wooded background and 8000 cd/m^2 in very flat and open country. In selecting the appropriate figure for lighting design, the finished approach structure must be envisaged. The luminance may be kept low by such expedients as a dark roadway with dark retaining walls or other structures, or by screening with trees. From the value of the luminance at this point the necessary luminance

Fig. 6.6 – (a) Portal of first Dartford Tunnel, showing lighting transition effected by canopy over to reduce intensity of sunlight and intensified artificial lighting within the portal. (b) Canopy at Liverpool portal of Mersey Kingsway Tunnel, spanning five traffic lanes.

in the tunnel entrance is usually taken to be between 1/15 and 1/30 but sometimes as high as 1/10. A very high level of artificial lighting of 1500 to 2000 lux is required to provide luminance of perhaps 100–200 cd/m^2, even with the assistance of a light coloured roadway and walls. In some tunnels canopies have been constructed extending from the portal and admitting reduced daylight to the roadway while excluding direct sunlight. This certainly has merits in economising on cost of lighting, but at a substantial construction cost. It is found difficult to keep the screens clean and efficient, and in some climates snow and formation of ice create hazards.

6.6.3 Tunnel interior
Within the tunnel, which for this purpose includes any canopy or similar extension, four lighting zones are recognised, namely:

1. Threshold
2. Transition
3. Interior
4. Exit

Within the *threshold zone*, the luminance may be kept constant as calculated for the entrance or may decrease slightly in parallel with the natural decrease in the external approach zone, maintaining a constant ratio, such as 1/15 at the chosen safe distance ahead. The threshold zone need not extend the whole safe distance from the actual entrance, but from a point a short way outside at which the entrance and its level of luminance dominates the field of view.

The *transition zone* immediately succeeds the threshold zone and within this the luminance is progressively reduced to the value adopted for the interior. The rate of reduction is precisely defined by a curve of adaptation, which may very approximately be represented by halving the luminance every 3 seconds.

For the *interior zone* the recommended minimum luminances range from 10 cd/m^2 in urban tunnels carrying heavy traffic to 3 cd/m^2 in very long tunnels in open country or where speed is restricted or traffic is low. The corresponding illuminance ranges from 175 to 50 lux for average reflectance of the road surface.

The *exit zone* in a one-way tunnel does not usually require extra lighting. In special cases, such as very long tunnels, or emergence to the very high luminance of snow or sea, a short transition zone of increasing luminance may be helpful.

It will be appreciated that the lighting in threshold and transition zones has to be varied throughout the day in proportion to the natural luminance at the origin, more particularly at twilight, but also where daylight fluctuates greatly between sun and storm. The control may be manual, or by time switching, but regulation by photometer sited outside the tunnel near its entrance is likely to be most satisfactory for major tunnels.

At night the tunnel lighting should be closely comparable with that in the approaches. If too bright, the exit becomes a black hole. If roads are generally unlit the luminance within the tunnel should be between 1 and 2 cd/m^2.

Short tunnels. The threshold and transition zones designed as described will occupy a length of 300 m or more from the entrance so that in a shorter one-way tunnel the exit portal will be reached before lighting levels have been reduced to the recommended minimum interior levels, and in two-way tunnels the critical length is doubled. It will be appreciated that the cost of such lighting is proportionately higher than for long tunnels.

It is difficult to generalise about the appropriate installations for such short tunnels down to the length where there is through visibility and in which artificial lighting, except at night, can be dispensed with. Each tunnel must be considered separately in relation to approaches, length, curves, vehicle speed, traffic density. Illuminated kerb lane markers may be helpful, or reflectors where use of headlights can be ensured.

6.6.4 Lighting installation

The most suitable systems are usually found to be fluorescent tubes, low pressure sodium or high pressure sodium. Uniformity of lighting, concentrated principally on road and walls, is desirable, and avoidance of flicker which can result from spacing out of lamps, both from the direct effect of the lamps themselves and their reflection in other vehicles. Uninterrupted longitudinal rows are usually the most suitable for heavy urban traffic with multiple rows where enhanced lighting is required. When lower intensities are appropriate alternate tubes may be switched off. Spacing exceeding 12 m is conducive to flicker.

The lighting circuits have to be designed to allow lighting to be adapted to varied daylight levels. It is also important that lighting should be so planned that the tunnel is not plunged into darkness if one circuit fails, and also with minimum emergency lighting against the contingency of major supply failure.

6.7 TRAFFIC MANAGEMENT

Management and control of traffic is found necessary in many tunnels if they are to be used to the best advantage with safety. In the absence of any control the tunnelled highway should conform fully to the dimensions and standards, including clearances, of the open highway and should impose no separate restrictions. Even at best, there are differences of environment likely to cause many drivers to slow down as they approach, unless they are very familiar with the route. Although clearances laterally and overhead may be the same as under bridges, the convergence in perspective makes them appear more restrictive. Lighting must be of a very high standard if it is not to cause hesitation by some drivers, and uncertainty for all.

Such ideal standards are rarely practicable in any but very short tunnels, whether from difficulties of construction or excessive cost. These physical limitations do in themselves exercise some restraint on traffic, although, as noted earlier, nominal road capacity may sometimes be substantially exceeded.

Various forms of restriction and control are applied in most major tunnels. Apart from imposition of tolls or stopping of vehicles at a national frontier, the primary reason for exercising control is safety. Other reasons are to ensure maximum flow, and to integrate the tunnel traffic flow with city traffic.

The ordinary safety regulations applied elsewhere are adopted in tunnels as found appropriate. Limits on height and width and maximum speed limits, which may be special to the tunnel, or general to the highway, are usual. Lane changing is frequently prohibited, often with restriction of heavy vehicles to a single lane. In two-way tunnels overtaking is usually prohibited. These are all controls of common application on highways and may not necessitate separate enforcement, but the consequences of any accident or breakdown is potentially so dangerous, and likely to be so disruptive to traffic, that special patrols are often employed, and also TV coverage to identify location and cause of congestion as early as possible.

Important loads exceeding regulation width or height are sometimes routed through tunnels when alternative routes are not practicable, but the closing of part or all of the tunnel to other traffic during the passage of the load is necessary. Extra height may be available in the middle of the roadway.

The safety limitations on traffic special to tunnels are in respect of ventilation and carriage of dangerous goods. Ventilation, discussed in detail in Chapter 9, is usually adequate, even in congested traffic, in busy tunnels, but there are important exceptions, where the traffic capacity of the tunnel is limited by the ventilating air available. Such is the case in the Mont Blanc tunnel, which, even after an increase, only provides for traffic at a rate of 800 veh/hr at a spacing of 100 m, although the road capacity is much greater. Other major Alpine tunnels are similarly limited.

Supervision of traffic and ventilation must be adequate to ensure safety if congestion does develop. Automatic control of fan output in relation to measurement of pollution levels is usually too slow.

6.7.1 Dangerous goods
Materials inherently dangerous by reason of their flammable, explosive or toxic nature are widely transported by road. Although accepted as a necessary hazard on the open road, they may be quite unacceptable in a tunnel or admitted only with special precautions. Their passage on the highway is not in itself hazardous, but the contents of the load may be released by collision damage or overturning, or by a fire in the vehicle. The risk that a collision or other accident will occur in a tunnel is no greater than outside, but the potential consequences in the confined space of a tunnel are in many cases quite disproportionate. They will

vary, depending on the characteristics of the tunnel and the traffic present in it. Regulations to be made and control to be applied should therefore be carefully considered to meet the particular circumstances and needs of each highway tunnel, and to contribute to its primary purpose of carrying highway traffic.

One of the important classes of dangerous goods is liquid fuel, whose transport in road tankers constitutes a substantial proportion of heavy vehicle traffic. The major danger from this and other flammable liquids is that of fire from spillage following collision. An added hazard, more particularly in sub-aqueous tunnels with their concave profile, is of the liquid flowing into the drainage system and sump, and possibly evaporating to form an explosive mixture. Such tankers are usually in charge of careful and skillful drivers, and in some tunnels are allowed through only at offpeak hours and when escorted by tunnel staff in a patrol vehicle.

The danger from explosives is obvious, but could be catastrophic in a subaqueous tunnel, or in one where structural damage might be followed by collapse. The use of alternative routes is to be preferred, but escorted passage through the tunnel with temporary exclusion of other traffic is possible. Toxic materials, more particularly in gaseous or liquid form, and materials becoming toxic in contact with water, also may be excluded, or may have to be passed under control.

In any such control system one of the greatest difficulties is to identify and intercept vehicles with dangerous loads. Where there is a toll barrier all vehicles can be inspected, but in many cases toll collection is concentrated at one end of the tunnel, half the vehicles having already passed through.

6.7.2 Means of traffic control
In a controlled tunnel the organisation will include any or all of the following:

1. Toll barriers
2. Traffic lights and illuminated signs
3. Staff patrol cars and breakdown vehicles
4. Closed circuit TV
5. Public address system
6. Instruments for monitoring carbon monoxide, visibility, numbers of vehicles, speed

6.7.2.1 Toll barriers
The collection of tolls in itself exercises a discipline on tunnel users but is something of an obstruction to free flow of traffic. When at full capacity, each traffic lane needs between two and three toll booths to ensure its full utilisation. A simple toll structure requiring payment with a single coin is ideal, and can operate with very little delay where toll collectors are equipped with suitable recording instruments and where drivers are familiar with the procedure. Tokens

Fig. 6.7 – (a) Wallasey approach to Mersey Kingsway Tunnel showing: (1) Access junctions. (2) Toll plaza with space for 12 toll booths. (3) Open cut approach to portals. (4) Ventilation towers on both sides of river.

Fig. 6.7(b) — Control room above toll booths.

or tickets can be helpful, but season tickets can be troublesome as they may be misused by drivers, or exploited by toll collectors, because they break the direct check between vehicles passed and tolls collected.

The passing of the toll booth is usually regulated by a traffic light which is operated by the toll collector's action in recording the receipt of the toll and the classification of the vehicle. It is usual for these records to be registered simultaneously at a central office. Automated toll collection has been tried, but has not yet been found entirely satisfactory. As compared with efficient manual collection from a regular user, no time is saved, and the driver requiring change must in any case use a manned entrance. A driver unfamiliar with the tunnel is likely to be delayed by the need to read the instructions about the use of the instrument and to find the correct amount of cash.

The space for 2½ lanes per tunnel plus the width of a toll booth plus safety clearances makes necessary a very wide area for the toll plaza. It may also be necessary to leave space for emergency access by fire engines or breakdown vehicles at a time when all traffic is halted and all toll lanes are congested. If all tolls are collected at one end of the tunnel space may be saved by making a number of the lanes and toll booths reversible, unless the peak flows occur in both directions simultaneously, which is most unlikely.

6.7.2.2 Traffic lights and illuminated signs

Within most tunnels traffic lights are undesirable. Their presence causes uncertainty to drivers, and they are unlikely to be of much use in emergency. The place for stop signals is in advance of the portal to exclude traffic from the tunnel when necessary.

Normal road signs cannot usually be accommodated within a tunnel and any instructions to drivers can best be given by special illuminated signs, operated when required. One which may be considered necessary for conditions of acute congestion, to minimise generation of fumes, is 'Stop. Switch off engine'.

Signs indicating fire points and emergency telephones may also be installed.

6.7.2.3 Patrols and breakdown vehicles

Car patrols are almost essential to supervise traffic and deal with emergencies. The duties may be carried out by traffic police but in many cases special tunnel staff are employed with police status and with a fleet of vehicles belonging to the tunnel. One of the duties of such patrols may be to escort dangerous loads through the tunnel.

Breakdown vehicles must be readily available, capable of clearing obstructions quickly. It is advisable to have them provided at both ends of the tunnel, as the scene of an accident may not be accessible from one end because the roadway is blocked with stopped vehicles.

It is necessary to ensure that the recovery vehicles can operate properly within the confines of the tunnel. For speed and efficiency it is advantageous

to have heavy recovery vehicles, light recovery vehicles, and for the patrol cars
to have some capacity for towing out broken down cars.

Fig. 6.7(c) – Heavy breakdown vehicles for use in tunnel. (*Mann Egerton &
Co. Ltd.*)

6.7.2.4 Closed circuit TV

TV cameras in the tunnel displaying as required in the control room can be of
great help in identifying and locating any incipient congestion or other trouble.
They should have the capability of showing any part of the tunnel and also the
traffic approaching the portals. A number of screens are likely to be kept on
permanent display with the capacity to switch on and direct any one of the
cameras. If there is a radio link with patrol cars—for which special devices are
necessary within the tunnel—fire fighting or other emergencies can be directed
from the control room.

6.7.2.5 A public address system

Such a system is only for use in emergencies. It could not normally be heard
by drivers in closed cars, and against traffic noise, but could be used to give
instructions and prevent panic in a fire.

6.7.2.6 Instrumentation

Levels of carbon monoxide must be monitored continuously to give guidance
on ventilation. Instruments should be installed at strategic points in the tunnel

air flow. Visibility, likewise, should be measured at a number of points. For both of these continuous recorders in the control room are required and some form of alarm to indicate critical levels.

The importance of knowing traffic flow, spacing and speed has been discussed. The use of a number of electromagnetic inductive loops under the carriageway coupled to a central computer allows these factors to be assessed and used. Older methods of counting by interception of a beam of light or by contact pads have not proved very reliable, if the aggregate difference between vehicles in and vehicles out is to be used to count the number of vehicles in the tunnel at any moment. Confusion arises from multiple axle vehicles and trailers and from vehicles not centrally in the lane.

6.8 FIRE PRECAUTIONS

Although serious fires in road tunnels are of rare occurrence the potential consequences are usually thought to make appropriate special precautions, more particularly in long tunnels carrying a heavy volume of traffic. Some of these requirements have already been noted above as an aspect of traffic management and the regulation of ventilating air is referred to in Chapter 9.

Two major examples of fires may be adduced:

The Holland tunnel in New York suffered from a major vehicle fire in 1949. A load of carbon disulphide in drums, which had entered contrary to regulations, caught fire and burned fiercely with emission of flame and thick smoke, which made close access impossible. No lives were lost but 66 were injured and 180 m of the lining was destroyed. The tunnel was closed for 2½ days.

In the Nihonzaka tunnel (opened in 1969) in Japan on the Tokyo–Nagoya highway there was a very serious fire in July 1979 with the loss of 6 lives. There are twin two-lane tunnels about 2000 m long. A heavy flow of traffic entered the tunnel when released after a long hold-up due to earlier collision. About 400 m into the tunnel there was a collision between two cars and four lorries, one of which was carrying liquid ether, which ignited. A sprinkler system operated but water supplies proved inadequate and the fire raged for 2 days, causing spalling of concrete from the crown with exposure of reinforcement, and burning out 160 cars. Regulations on transit of dangerous goods were not applicable to this tunnel.

In major tunnels it is usual to provide fire points at intervals of 50 or 100 m along the tunnel, at which are installed

1. press-button fire alarms and emergency telephones connected to the control centre, and if possible to the fire brigade;

2. hose reels carrying 50 m or so of 37 mm hose;

3. portable chemical extinguishers, and

4. outlets complying with fire brigade standards connected to a high pressure
    fire main of about 100 mm diameter or more.
    These fire points are usually on the walkway and should be readily access-
    ible to any driver in trouble. They should be clearly marked with illuminated
    signs.

The dangers to life consequent on a fire are caused by heat, smoke, oxygen
deficiency, carbon monoxide. The importance of a fresh air supply and an
extraction system in combating these is obvious, except that continued combustion depends on air supply, which should therefore be carefully regulated
if practicable.

Heat is a direct danger to life including the radiant heat from burning gases
flowing under the ceiling. It also attacks and impairs the strength of structural
materials which are not themselves combustible. In a major fire reinforced
concrete may suffer spalling of concrete followed by loss of strength in the
reinforcing steel. Smoke destroys visibility and may also contain toxic constituents; its extraction at roof level assists escape of persons at low level.
Oxygen deficiency and carbon monoxide may develop in a major fire but
more particularly in the absence of any fresh air supply.

Aspects of fire risks to be considered in design include, in addition to
ventilation:

1. escape routes, possibly by cross passages to an adjacent tunnel, of particular
    importance in very long tunnels;
2. provision of emergency lighting adequately protected against fire;
3. choice of lining materials, particularly in ceilings and ceiling ducts, which
    will neither contribute to combustion, nor give off toxic fumes, nor collapse;
4. appropriate arrangements to drain and trap burning fuel spilling on the
    roadway;
5. protection of cables and wiring wherever potentially vulnerable.

6.9 PIPES AND CABLES

A tunnel itself requires a range of services supplied by pipes and cables. In
addition, as already mentioned, public authorities may make arrangements for
their mains to be routed through the tunnel, more especially in a subaqueous
tunnel.

6.9.1 Water

A fire main in a tunnel is a normal provision. Water may also be required for
washing down and these services may be separate, or combined. Trunk mains
of larger diameter may also be installed to link the public systems at the two
ends. Those water pipes can usually be accommodated without difficulty or
danger, if safeguards are provided against the risk of flooding from a burst.

Frost protection may be necessary if the water pipe is exposed to fresh air being drawn in for ventilation.

6.9.2 Gas

There is no normal requirement for a gas supply to the tunnel. High pressure trunk mains are not usually acceptable, in particular because of the risk of explosion following a leak or burst, and the difficulty of providing adequate safeguards.

6.9.3 Power cables

Tunnel supplies for lighting must be carried through the tunnel; those for ventilation and pumping may also be so routed together with linking power cables to allow standby supplies to be available from either end. These will usually be in ducts giving some protection against mechanical damage and against fire. Spare ducts may be provided for subsequent developments or for replacements.

Public supply cables forming part of the grid may also be sited in the tunnel. If carrying heavy currents they may require to be cooled either by

Fig. 6.8 – Tunnel service cables behind secondary lining in second Dartford tunnel. (*B.I.C.C.*)

use of the ventilation air supply or by water cooling or other means. Automatic protection can be given against explosive cable rupture. Mechanical protection against damage by collision may be ensured by appropriate siting.

6.9.4 Communication cables
These are likely to be required for transmission of instrument readings and control relays and for telephones and alarms, also for any TV and public address systems. They will usually be carried in ducts or conduits, with accessible junction and distribution boxes at intervals. Watertightness is important and may be difficult to achieve.

BIBLIOGRAPHY

There is some difficulty in presenting a bibliography for this chapter in that most construction aspects of highway tunnels are not peculiar to their function, and have thus been dealt with and included in other chapters, including in particular Chapter 1 — Cut-and-cover, and Chapter 2 — Submerged Tunnels, to which reference should be made. The requirement for ventilation is so complex and specialised as to be treated separately in Chapter 9.

It has, however, been thought of interest to list here the whole sequence of U.K. subaqueous highway tunnels, and a few others, from 1897 to the present day, as they have been recorded in useful detail. For worldwide practice generally, the P.I.A.R.C. reports apply a necessary corrective to this insular perspective. References to tunnel lighting are also given.

Hay, D. and Fitzmaurice, M., The Blackwall Tunnel, *Min. Proc. Instn Civ. Engrs*, 1897, **130**.

Tabor, E. H., The Rotherhithe tunnel, *ibid*, 1908, **175**.

Anderson, D., The construction of the Mersey Tunnel, *J. Instn Civ. Engrs*, 1936, **2** (Apr.).

Van Bruggen, J. P., The road tunnel under the River Maas at Rotterdam, *Engng*, 1940, **150** (Aug. 9 and 30, Sept. 27).

Kell, J., The Dartford tunnel, *Proc. Instn Civ. Engrs*, 1963, **24** (Mar.); *see also* discussion, 1963, **26** (Dec.).

Schreuder, D. A., *The lighting of vehicular tunnels*, Philips, 1964.

Granter, E., Park Lane improvement scheme: design and construction.

Rayfield, F. A. and Clayton, A. J. H., ditto: planning and traffic engineering, *Proc. Instn Civ. Engrs*, 1964, **29** (Oct.); *see also* discussion, 1966, **33** (Mar.) and **35** (Dec.).

Morgan, H. D. *et al.*, Clyde tunnel: design, construction and tunnel services.

Haxton, A. F. and Whyte, H. E., ditto: constructional problems, *ibid*, 1965, **30** (Feb.); *see also* discussion, 1967, **37** (July).

Kell. J. and Ridley, G., Blackwall tunnel duplication, *ibid*, 1966, **35** (Oct.); *see also* discussion, 1967, **37** (July).

Road Tunnels Committee:
  Report and *Documentation and Studies*, 13th World Congress, Tokyo, 1967;
  Report and *Documentation Digest*, 14th World Congress, Prague, 1971;
  Report and *Documentation Digest*, 15th World Congress, Mexico, 1975;
  Report, 16th World Congress, Vienna, 1979;
  Permanent International Association of Road Congresses, Paris.

Prosser, J. R. and Grant, P. A. St. C., The Tyne tunnel: the planning of the scheme.

Falkiner, R. H. and Tough, S. G., ditto: the construction of the main tunnel, *Proc. Instn Civ. Engrs*, 1968, **39** (Feb.); *see also* discussion, 1968, **41** (Nov.).

Lyons, A. C. and Schofield, J., The Great Charles Street tunnel, *Tunnels and Tunnelling*, 1969, 1 (May).

Margason, G. and Pocock, R. G., *A preliminary study of the cost of tunnel construction, Laboratory Report 326*, Transport and Road Research Laboratory, Crowthorne, 1970.

Parry, R. R. and Thornton, G. D., Construction of the Monmouth tunnels in soft rock, *Proc. Instn Civ. Engrs*, 1970, **47** (Sept.); *see also* discussion, 1971, **49** (July).

Lyons, D. J., Trends in research on motorway design and use in *Motorways in Britain today and tomorrow*, London, 1971.

Brown, C. D., Homes and highways, *Consult. Engr*, 1971, **35** (Nov.).

Muir Wood, A. M., Tunnels for roads and motorways, *Q.J. Engng Geol.*, 1972, **5** (1 and 2).

Megaw, T. M. *et al.*, The Mersey Kingsway tunnel (3 papers), *Proc. Instn Civ. Engrs*, 1972, **51** (Mar.); *see also* discussion, 1973, **54** (May).

O'Reilly, M. P. and Munton, A. P., Prospects of urban highways in tunnel, in *Transportation Engineering 1972*, London, 1973.

Sir William Halcrow & Partners and Mott, Hay & Anderson, *Report on Roads in Tunnel for the Greater London Council*, G.L.C., 1973.

International Recommendations for Tunnel Lighting, Pub. no. 26, Commission Internationale de L'Eclairage, Paris, 1973.

Muir Wood, A. M. and Sharman, F. A., Tunnelling as a social benefit, in *Transportation and Environment: policies, plans and practice*, Southampton, 1973.

Symposium on Tunnel Lighting, London, 1974, *Lighting Research and Technology*, 1975, **7** (2).

Shutter, G. B. and Bell, G. A., Design and construction of the second Dartford tunnel, in *Tunnelling '79*, London, 1979.

7

Metro Tunnels

7.1 CHARACTERISTICS

The term 'metro' is being increasingly used to distinguish railway systems specially adapted to cities and their immediate environs. The terms underground, U-bahn, subway, rapid transit, S-bahn, used in many cities are less distinctive. It is of interest that the earliest such system was London's Metropolitan Railway, followed by the Paris Metro.

The most important characteristic of a metro is the ability to transport rapidly large numbers of people on an exclusive right of way, free from any interruption by other types of traffic. Frequent, predictable and reliable services are of great value to the passengers, of whom a majority travel regularly between home and work, with many also travelling in and out of the centre for education, shopping, entertainment, and also within the inner city area.

Metros are to be distinguished functionally from main line and suburban railways by their use for shorter and more frequent journeys although no hard and fast line of demarcation is possible. A possible definition of a 'metro' is: an urban passenger railway with stations at frequent intervals to provide convenient and rapid transport over relatively short distances within a city and its environs.

Metro systems are almost invariably required to meet the needs of existing cities, whose centres are so closely built up already that surface railways or elevated railways are quite impracticable and tunnelling becomes essential. Tunnelling, either by cut-and-cover under existing streets or by boring under streets and buildings, minimises interference with existing traffic capacity, demolition of buildings and visual intrusion.

Tunnelling costs are necessarily high and therefore it is usual to tunnel only in congested areas and under high ground, and to run on the surface or as an elevated railway wherever adequate space is more readily available, possibly over or alongside an existing highway.

Correct siting of metro stations at close spacing in the inner city is an obvious necessity if maximum traffic is to be attracted, although in suburban areas stations must be more widely spaced to avoid the delays and costs inherent in frequent stops. A balance has to be established between these delays and costs and the extra traffic attracted. In central tunnelled sections the device of constructing double-ended stations can increase substantially the catchment area of a single station. Entrances, booking offices, and access stairs are provided at both ends of the station platforms and may therefore be more than 200 m apart which is often a significant proportion of the distance between stations.

Before planning a route in any detail it is essential to have an assessment of the number of passengers to be expected and their origins and destinations, and more particularly the number during a 15-minute peak. From this information alternative proposals for possible routes can be prepared and a single route selected for detailed study of its traffic potential and engineering aspects.

The enclosed nature of metro systems with their own right of way makes possible automation of many aspects, thereby reducing the labour intensive character of public transport generally. Train operation, signalling and ticket issue and control can all be regulated automatically and largely centralised, but the fact that the system handles human beings limits the extent to which machinery can function satisfactorily in relation to passenger convenience and safety and in meeting emergencies.

7.2 COMPULSORY POWERS

For the construction of a metro system in a city compulsory powers are usually necessary to acquire land and rights of way, and for operation and maintenance. Provision must also be made for appropriate compensation, the assessment of which should not be allowed to delay planning and construction.

In the United Kingdom a Private Act of Parliament is normally obtained for the purpose by the promoters, describing the project in sufficient, but not excessive, detail and prescribing the centre line within defined limits of deviation. Other forms of authorisation will be more appropriate elsewhere but the nature and limitations of the powers to be exercised by the promoters must be essentially similar.

The right to compensation of property owners adversely affected by construction or operation requires to be defined while the promoters must be given rights to enter property for surveys, for trial borings and soil investigation, and also for temporary and permanent underpinning and strengthening. A time limit is usually defined within which the rights must be exercised, but with possibility of extension. Claims for injurious affection by construction or operation are also normally limited in time.

On the other side of the balance it is usual that the provision of good rapid transit enhances property values where it serves a residential area, and also improves business and shopping areas. There may also be substantially reduced congestion in streets. Such benefits are dispersed throughout the community and are not received by the promoters to offset against the necessary items of compensation; it is usual and proper therefore that the central authority should pay a substantial part of the costs incurred. Greatly enhanced land values followed the construction of the Toronto system, but similar effects were not experienced in the early years of the BART system in California.

The negotiations with those affected by a new line are likely to be far more difficult for cut-and-cover construction than for a deep tunnel. Objections sometimes are given disproportionate publicity; the promoters have to justify their proposals in much greater detail.

The special problems to be examined include:

1. the establishment of a legal right of way;
2. siting of stations to suit prospective passengers;
3. layout of running tunnels linking stations within acceptable limits of gradient and curvature;
4. avoidance of underground obstructions such as deep foundations and existing tunnels and shafts;
5. adaptation to geological conditions;
6. maintaining appropriate distance from establishments particularly sensitive to vibrations such as laboratories, large hospitals or concert halls.

Each of these aspects influences the others and the structure finally adopted must be a compromise on the basis of estimates of feasibility and cost of possible alternatives.

Access necessary for construction must be determined as a compromise between conflicting ideals. Working sites are essential but are hard to find in densely built-up areas, and are often subject to restrictions in respect of noise, loss of amenity, access by heavy vehicles, or prospective building construction. Permanent acquisition of sites for stations, or for ventilation or electrical supply may provide many of the necessary points of access. For cut-and-cover construction the obstacles are much more severe: streets must be occupied for a long period, excavated and ultimately restored or, if streets are not followed, buildings must be cut through or demolished with little respect for property boundaries.

When the layout of the line and its stations has been settled, ventilation and associated problems of draught relief need to be considered. The extent and complexity of the system necessary depend on the climate of the city and the standards adopted, but siting of shafts and provision for fans should be an integral part of metro design and necessary land should be acquired before construction of the line begins.

Survey to a high standard of accuracy on and under the surface and including location of buried services, is an essential preliminary.

Planning must not overlook requirements for inspection and maintenance when the system becomes operational. The limited clearances in tunnels and the short intervals between trains, restrict these operations to those hours when train services can be suspended and current cut off from a live rail or other conductor. Except where a twenty-four hour service is maintained, a few hours are usually available each night, but such operations are necessarily very costly and should be kept to a minimum by careful design and choice of the most suitable systems. It is even more difficult and costly if all-night services are run.

7.3 GENERAL PATTERN OF PLANNING

The planning of a metro system is extremely complex and involves many different and conflicting interests. It is one element only of a city's long term planning but one which will affect the future growth and living pattern of that city. Such broad plans have necessarily and properly political aspects, but the specific selection of a route and its detailed design is properly a matter of engineering judgement. The intention here is to outline such aspects as are directly relevant to construction and use of tunnels.

Decisions in principle should be embodied as early as practicable into a full set of design criteria, progressively developed into precise specifications governing such matters as methods of construction, types of rolling stock and other plant and equipment, operating speeds, track geometry, tolerances, air temperatures, rates of air flow and very many others. One very broad decision to be taken early is the level of mechanical complexity and automation to be adopted for operation. This will be influenced by the ultimate availability, after training, of suitable operating and maintenance staff.

Capital cost is inevitably high, more especially in congested areas, and expenditure on construction must be justified by providing a transport system serving the needs and wants of a large number of passengers. In modern cities very many people have their homes as far as practicable from the centre, and travel to and from the centre for business and pleasure. This establishes a demand for transport on radial routes and success in meeting this demand conveniently and economically encourages more people to adopt this pattern of living and travel. Therefore both an analysis of existing wants and predictions of growth require careful study in planning a route. The principal routes are often apparent along corridors determined by the topography. Sometimes a line may deliberately be so laid out as to encourage development of a particular area, while catering also for existing demand from other areas.

A single radial route is not satisfactory as it involves reversing trains at a central terminal but a diametrical route crossing the centre, or a combination of two radial routes with through running is very much better. The two radial lines may even be nearly parallel but connected by a U-loop. The first lines of the Toronto Subway embodied this pattern; no lines can run out to the south because of Toronto's situation on the shore line of Lake Ontario.

A system comprising separate and relatively short lines has the advantage that any delays are confined to one line and thus affect a minimum number of passengers.

Radial routes are thus the normal pattern for city transport, but circular routes have also been adopted. The earliest metros in London were linked to form the 'Inner Circle' in 1884; its special function was to connect the main line termini which, for reasons of congestion, were widely separated and were sited round the perimeter of the city's core. Glasgow also in 1896 adopted a circular route, which has not been added to subsequently. Another example of a circle route is in Moscow where it is in addition to the numerous radial lines.

Circular systems are far less common than radial lines and are only justified in special circumstances. Where both coexist convenient interchange for passengers is essential but interworking is usually found to be operationally undesirable, because delays on one radial line can disrupt the circle service and in turn all the other radial lines. In addition, with interworking the train capacity of the circle line limits the use of the radial lines unless each has its own separate track round the circle.

An outer circle line is rare and is more difficult to justify, particularly so if tunnelling has to be considered. It would serve as a bypass to the city, or for communication between various suburbs or satellite towns but it is most unlikely to attract traffic of sufficient intensity to justify the capital and operating costs.

In addition to the running route a rolling stock depot is required for stabling of trains and day-to-day maintenance. Provision must also be made for major overhauls, whether at the same site or elsewhere. A large area of land is necessary which will obviously be most economically acquired as a surface site remote from the city centre, towards the outer end of the line.

The systems considered above are self-contained units, but there may also be direct links through a city centre between existing main or suburban lines having many of the characteristics of metros. The new Melbourne Underground Loop is such a system, comprising an inner circle carrying services from suburban lines and providing the advantages of frequent stations in the centre serving all radial lines, and the operating benefits of through running of trains without the need for reversals and stabling. The disadvantages noted above of interference between radial and circle services are avoided by the provision of four loop tracks each in a separate tunnel large enough to carry standard rolling stock, and also possible double-decker suburban stock.

On a smaller scale the new Liverpool 'Loop' line extends the cross-Mersey line as a one-way loop providing passenger interchange at the main line stations and eliminating reversal of trains. It is complemented by the 'Link' line connecting two local lines across the city and providing interchange with the 'Loop'.

The choices between cut-and-cover and deep bored tunnels are discussed in more detail below, as is the question of size of rolling stock to be accommodated.

Fig. 7.1 – King's Cross Underground Station in 1868.

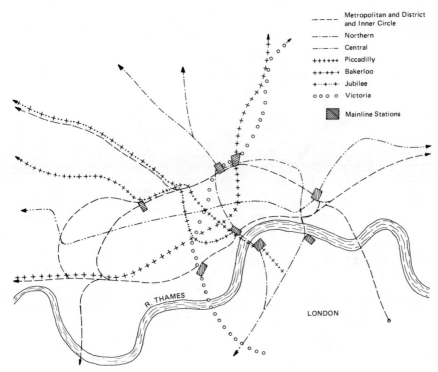

Fig. 7.2 – Line diagrams of metros: (a) London,

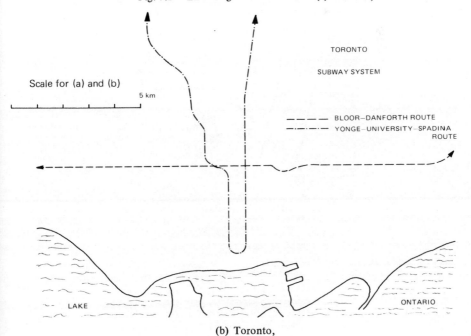

(b) Toronto,

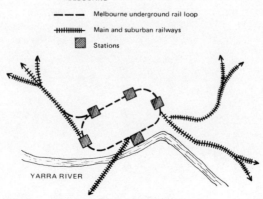

(c) Melbourne,

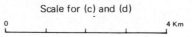

(d) Liverpool.

Fig. 7.3 – Opening of London Central Line, July 1900. (*London Transport*)

7.3.1 Control centre

A central control establishment is now becoming of major importance for the efficient operation of a modern metro. The primary function is the supervision and regulation of train services, for which purpose full information must be continuously furnished and displayed.

Other mechanical functions which can usually be centralised with advantage are control of the various electric current supplies, and the remote operation of plant and equipment such as escalators, control gates, ventilation, pumping, emergency flood gates, etc.

Closed circuit TV and other links can provide essential information from stations, both by viewing passenger movements and by transmitting particulars relevant to operation of equipment. Such a control centre is also the proper focus for relevant information from outside sources—weather forecasts, flood warnings, crowd movements related to sporting or other events, road accidents and diversions.

7.3.2 Establishment of routes

Having decided on the areas to be served by a new line its route must be laid out in detail with consideration of the necessary rights of way and land acquisition, as well as the engineering possibilities.

The balancing of costs against benefits is of major importance in justifying the expenditure to be incurred although neither costs nor benefits can be quantified with any precision. Costs of land acquisition and engineering construction, equipment and maintenance can be estimated reasonably well in financial terms, as can passenger use and income from fares, but other large factors such as environmental intrusion, disruption during construction, relief of road congestion, enhancement of land values, etc., cannot be evaluated with any precision, and are not likely to be of any direct benefit to the immediate promoter unless he is also a property owner, or a transport operator, public or private. In choosing the exact alignment, both in plan and in level, more detailed examination of the comparative costs and benefits of alternatives is necessary.

The two main elements of direct cost are land acquisition and construction. Acquisition of a right of way comes first. There may exist a surface right of way which can be used, following a major road or within the boundaries of a railway. Even in the absence of such a right of way over part or the whole of the line, it may be cheaper to acquire the necessary new land for a surface line than to tunnel. This is unlikely to apply in the central area of a city where tunnelling on a substantial scale is to be expected. Here too a right of way under the ground is required, and a line beneath existing streets has many advantages, provided sharp curves can be avoided. In many cities such street routes are natural traffic arteries and thus meet readily the needs of passengers. In the past, however, streets have sometimes been too closely followed and the railway has been constructed with undesirably severe curves, and even reverse curves. Acquisition of necessary land or rights to pass under private property at such points is likely to be fully justified. There are also routes within a city where the best line departs completely from the surface street pattern.

Siting of stations is a most important aspect of detail planning. Experience in London and elsewhere has shown that a station sited where inconvenient to the public will significantly fail to attract passengers to the metro. A station should be sited as close as possible to a major traffic focus, but even if the ideal site cannot be achieved, it may be worth while to provide pedestrian subway access to the station, incorporating escalators or horizontal conveyors where they can appreciably improve the time and effort of getting to the trains. It is very important that entrances to stations should be sited in the best places, at street level, and attractively designed and clearly signed; back street entrances of dull appearance and not readily identifiable will do much to discourage passengers.

The integration of metro stations in a city as part of a larger complex of public buildings, shops and surface transport is a wider subject discussed in more detail later in this chapter.

Stations providing interchange with other metro lines or with main line railways require particular consideration in planning. Changes in the course of a journey are a deterrent to use of the system, which must be minimised. The time and convenience of passengers needs to be carefully studied, keeping distances and differences in level as small as possible. To be able to change trains by simply crossing the width of a platform is obviously excellent, and considerable ingenuity has sometimes been exercised in arranging tracks to provide this layout. Where lines cross at right angles a substantial difference of level is inevitable and escalators may be provided. It is worth noting that where both lines operate on a through ticket the interchange passages should be entirely within the 'paid' area to avoid unnecessary checking.

At stations on the outer sections interchange with buses and private cars may be an important problem involving car parking and waiting facilities, but these are unlikely to involve tunnelling, except for pedestrian access subways.

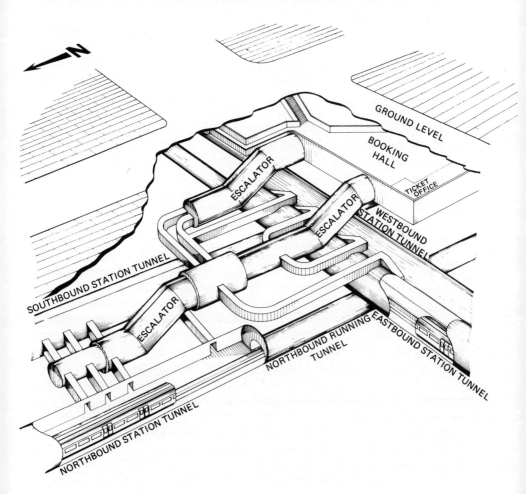

Fig. 7.4– Station structures where two lines cross. (*After London Transport.*)

The design of a station will be governed by the forecasts of peak passenger flows upon which will be decided the area of ticket hall, circulating areas, numbers of escalators or lifts, widths of stairs and passages and the access provided to platforms. Booking hall size and shape will be adapted to the ticket vending and checking system proposed. Platform lengths must always accommodate the longest train required in peak periods, and this length must be provided even at the less-used stations and near the outer limits of a line.

Train movements in the tunnels cause air flows in a station which may be found objectionably high, particularly in passages, escalators or ticket halls, unless the station is designed to accommodate maximum flows by appropriate layout of all passages and adequate cross sections, assisted where appropriate by draught relief shafts.

Other aspects of station planning may include space for shops, kiosks and facilities for the public, as well as staff accommodation, all located to avoid as far as possible any interference with flow of passengers. Arrangements must also be made for handling of moneys and for removal of litter and refuse generally.

7.3.3 Information

The information necessary for planning and constructing tunnels is discussed in Volume 1, but some aspects of particular importance to metro construction may be emphasised. As the work is typically in city centres and heavily built-up areas, existing records may be expected to provide much detail of structure and buried services. The older the city and its buildings the less easy it is likely to be to get access to full and accurate records of foundations and other works below the surface.

Any other tunnels, in use or abandoned, may need to be surveyed with some precision if unnecessarily wide margins are to be avoided in fixing the line and level of the new tunnels. Deep basements, piled foundations, wells, including abandoned shafts, rivers and streams, may all form obstacles. Sewers, water mains, gas mains, electricity cables, telephones and other less usual utilities such as hydraulic mains and pneumatic tubes may all be present. For deep tunnelling they are unlikely to be serious obstacles except at station sites, although sewers may sometimes be very deep, and some water supplies are run in deep tunnels. For cut-and-cover all the services must be located as exactly as possible with a view to their protection and support, or diversion. Proposals for future development in the area which may affect the stability of the metro tunnel may sometimes be anticipated by strengthening the tunnel structure locally, or by constructing in advance suitable foundations.

For tunnelling under a river very full information is necessary on the depth, as existing and as it may be affected by any dredging, or by scour in floods or attributable to river training works. Available records of tidal levels and flood levels should be studied and continuing records preserved unless adequately maintained elsewhere. Piled structures for wharves and jetties and for bridge piers need to be investigated and, of course, any other bridge foundations and river walls.

7.3.3.1 Geotechnical investigations

General requirements for tunnelling are described in Volume 1, comprising initially study of existing information in maps, borehole records and records of previous construction, followed by preparation of a scheme for investigation of possible routes by boreholes, trial pits and shafts, and by geotechnical methods.

While existing records are likely to be extensive in a heavily built-up area, much of the information may be irrelevant to tunnelling, because it usually is related principally to the upper layers of the soil above tunnelling depth, and tunnel excavation problems are not considered.

The problem of obtaining sufficient suitable sites for boreholes is often difficult. Selection of positions may unfortunately be governed more by availability than by tunnelling requirements. In this respect tunnelling under streets has advantages, more especially cut-and-cover construction, because extensive surface occupation is implicit. Opportunities of boring on temporarily vacant sites sometimes occur briefly, and it is helpful if those planning the metro are able to take advantage of them quickly even if the full scheme has not been developed.

Parks and open spaces may offer favourable sites for borings, but severe restrictions on such use are often imposed.

The apparent merits of geotechnical investigations—electrical, magnetic, sonic and other techniques—are at a special disadvantage in urban conditions because of the presence of pipes, cables and other buried artefacts.

7.3.4 Type of construction and programme

While the selection of a route is primarily dependent on traffic needs, to meet which appropriate engineering techniques must be employed, the precise siting and detail layout of stations will differ according to whether bored tunnels, cut-and-cover, or surface or elevated tracks are adopted. The latter are rarely suitable near city centres and are not further considered in any detail.

The first possibility likely to be examined is that of cut-and-cover construction. A major operating advantage of relatively shallow stations lies in easier access from the surface, which is of great importance in attracting passengers to a system providing rapid transit, especially over short distances. The time taken in descending from the street to the station platform, and the waiting time determined by frequency of service may be critical in deciding whether the metro or other surface transport is used for a particular journey. Likewise, on journeys involving interchange, the time and effort necessary to change trains should be kept to a minimum.

There are two substantial disadvantages in shallow construction by cut-and-cover. First is the disruption inevitable during construction. Streets must be opened up for long periods with extensive disturbance of street traffic. Vehicle access to adjoining properties can only be maintained with difficulty and they are subject to inevitable noise and dust and other loss of amenity, often resulting in loss of business. Where streets are not followed extensive demolition of property is necessary or very costly underpinning.

Second is the necessity to pass under buried services—pipes, cables, sewers—and possibly under rivers and existing tunnels, which may force down the railway to a depth at which trench excavation is more expensive than bored tunnelling. This last factor is of increasing importance as a city elaborates its metro system with a number of crossing lines.

Although bored tunnelling appears initially the more costly method, the indirect savings in reduced disruption and loss of amenity and in reduced costs of property acquisition will frequently make it the preferred alternative. There is, of course, the handicap of deeper stations requiring longer escalators and therefore increasing the time for access by passengers.

In considering bored tunnelling the choice of the most suitable stratum will be important in deciding on depth. Information is particularly necessary on water table levels, and the possibility of tunnelling in impermeable strata. Depending on distance between stations and acceptable gradients, stations may be built at relatively shallower depths with the running tunnels descending and rising again between stations.

Another important consequence of the choice between cut-and-cover and bored tunnel is that of size and cross-section. Generally, a trench lends itself to a rectangular cross-section and can be made reasonably spacious without disproportionate increase in cost, whereas in a bored tunnel of circular or horseshoe section, extra space is disproportionately expensive.

7.3.4.1 Construction sites

At an early stage of planning it becomes important to find possible construction sites from which work can be executed economically. In cut-and-cover the whole line constitutes a working site progressively, but larger areas are required beyond the mere width of the tunnel for assembly, operation and maintenance of plant and storage and handling of materials. For bored tunnels similar areas are likewise essential with additional requirements of access to the tunnelling by shafts, pits or adits.

Although station sites are useful as working sites for tunnelling when available, the construction of the stations often cannot be subordinated to the tunnelling operations and other areas must be found for tunnelling operations.

Ventilation shafts, cable shafts and emergency exits may provide satisfactory temporary construction sites, and therefore should be determined as early as possible with this use in mind and with the intention of utilising working shafts for permanent requirements.

The spacing of such working shafts will depend on, and in turn influence, the depth and type of tunnel to be built, the use of a tunnelling shield or machine and the lengths of the drives; all being related also to the nature of the ground. The maximum practicable working area should be provided immediately adjacent to each shaft, although some of the requirement can be met by provision of a second area in the vicinity.

It is generally found that the construction and equipment of stations takes longer to complete than does the connecting tunnelling and it may therefore be appropriate to begin the building of stations first. Nevertheless, the installation of specialised railway equipment in the running tunnels can be a lengthy process and the time required should not be underestimated. (See Figs. 7.5a and b).

7.3.4.2 Staged construction

Where a long line of new railway is to be built it is advantageous to divide it into a number of sections which can be opened to traffic successively at intervals of a year or less. There are benefits both in construction and in operation. Contracts can be let progressively, avoiding peak loading of the construction industry and providing continuity of work over a longer period for those with special skills, while applying the experience gained on one section to the building of the next. Operationally, the staff responsible for bringing each section into use can be concentrated more efficiently, and likewise make use of experience gained. Electrical trackside and other equipment, liable to deteriorate in unavoidably humid conditions, can be operational more quickly.

If such staged opening is planned the early acquisition of a site for the rolling stock depot will be most valuable so that it may be used as the main depot for installation of track and other equipment in the tunnels. Where the depot is remote from the city centre this implies the opening first of an outer section of the line which again has the advantage of providing a training ground where the traffic is less intense than in the centre. A requirement of each stage is the provision of temporary facilities for reversing trains at the end of each section, and the stages may well be fixed by positions where an emergency crossover is a feature of the permanent design and can be utilised.

The staged opening affects all sections and aspects of the project: survey, acquisition of land, preparation and letting of contracts, supervision of the works, manufacture and supply of rolling stock and other equipment, and training of railway staff in operation and maintenance.

7.3.5 Track geometry

The basic geometry of a metro railway in terms of curves and gradients determines very closely the alignment of tunnels and of stations. Adequate margins of safety for stability of trains must be maintained, with curves designed to provide a comfortable ride for passengers.

Fig. 7.5(a) — Construction site for Victoria Line: — During construction.

Fig. 7.5(b) — Construction site for Victoria Line: — As subsequently restored.

The horizontal geometry is initially considered as comprised of straights and circular curves, which should have the maximum practicable radius. The curves are superelevated to match a specified design speed, usually somewhat less than the maximum. Transition curves are then inserted, modifying a length of curve and straight, so as to build up the curvature and superelevation at a comfortable rate. Unless sufficient length is available, adequate superelevation is not attained and speed restriction becomes necessary, although some deficiency from the calculated balance at maximum speed is acceptable.

Curves of small radius must be avoided, not only because of the long transitions required but also because of the excessive cant needed if normal maximum speed is to be maintained and because of the overhang of coaches on such curves, encroaching on tunnel clearances or making necessary a larger diameter of tunnel.

In some existing systems radii below 150 m have been employed but 400 m is a desirable standard as a minimum for new construction, although radii down to 200 m have been adopted for important systems in recent years.

Vertical alignment is regulated by the ruling gradient, minimum drainage gradient, station and sidings gradients and vertical curves. The maximum acceptable gradient for a long slope is limited by the motive power of the trains. Very steep gradients may be necessary in hilly cities, as for example in Montreal, but in less severe conditions 3% or 4% may be adequate as a maximum.

In order to ensure effective drainage, it is advisable that no length of tunnel between stations should be quite level. It is sometimes necessary to provide a low spot for a drainage sump and pumps. At stations it is desirable for safety that the train should not roll under gravity even with the brakes off. With modern stock this requires a level track, although previously a small gradient has been considered acceptable. Similar considerations apply to sidings, and certainly in no case should a siding fall towards a main line.

Vertical curves of large radius are required at changes of gradient, which should not be sited in the vicinity of crossovers because it is not possible to maintain satisfactorily switch blades in vertical curves.

Because stations are relatively closely spaced it is important that trains have braking and acceleration rates as high as is consistent with passenger comfort. The trains can be assisted by provision of a rising gradient approaching a station and by a falling gradient leaving the station; this constitutes a so-called hump profile. Some economy of current consumption and brake wear is also achieved.

7.3.6 Power supplies

Electrical power is required in a tunnelled metro system for traction, lighting, ventilation, pumping, escalators or lifts, and for the signalling and control systems as well as for various services in booking offices, staff accommodation and elsewhere.

Traction power is most usually DC, which has been found suitable in relation to the limited route mileage, generally less than 200 route miles. The widely used conductor rail system is generally DC within the range 600 V to 1000 V, while for an overhead DC system about 1500 V is common, and for an overhead AC system 6 kV to 25 kV. The driver of any train should be able in emergency to cut off the traction power supply in the section.

Station and tunnel lighting is usually fluorescent and may be supplied at 240 V AC through a local transformer substation. Emergency lighting from a separate energy source is essential. Ventilation and pumping, probably with remote or automatic control, may require a 240 V or 415 V supply between stations. Escalators and lifts demand substantial and reliable power. If duplication of their supply requires the running of power cables through the tunnels a high voltage AC connection is much to be preferred because of the fire risk from a cable carrying a heavy current at lower voltage.

Supply for the signal system must be available along the whole length of the line. It is preferable that this be a separate system so connected into substations that supply is maintained when any one substation is shut down. Voltage of about 600 is suitable and frequency of 50 Hz is acceptable although there is an advantage in having an independent frequency whereby interference is limited.

7.4 DESIGN

There is not a well defined boundary between planning and design; each reacts on the other, but broadly it may be suggested that engineering planning is done at scales smaller than 1 in 1000, while engineering design takes over at scales of 1 in 250, not ignoring the smaller scale layout plans.

The functional design of the various elements—structures, machines, equipment—is very closely restricted by the limitations of space in all three dimensions and precise geometrical calculation pervades the whole subject.

If existing rolling stock has to be accommodated, that must be the starting point for consideration of tunnel size, gradients, curvature and provision of appropriate track and trackside equipment. It is obvious that specially designed stock adapted to the proposed tunnel can provide great economies and advantages, provided that there is no requirement to accept other rolling stock from connecting lines. It is in bored tunnels of circular section that the disadvantages of standard railway stock are most apparent; carriages are usually rectangular in form and, in the circular tunnel which provides adequate clearance at the corners of a rectangle, space overhead and at the sides is likely to be larger than necessary. London's 'Tubes', including recent and current projects, exemplify great economy in the proportion between passenger carriage area and excavated cross section. Early standardisation of tunnel diameter as 12′6″ (3.81 m) external and 11′8¼″ (3.56 m) internal, increased progressively to 3.81 m internal, has governed design and development of rolling stock. Most new systems in other cities are of larger dimensions, as for example Toronto, Washington, Hong Kong and Brussels, either because the rolling stock was initially developed to suit cut-and-cover sections or because greater importance was placed on the spaciousness inside the cars, especially in hot climates. Some authorities insist on the ability to have staff in the tunnels during service hours and this naturally makes large clearances necessary.

In the tunnel design the aim must be to provide a structure requiring no major maintenance. Its alignment, horizontal and vertical, must be the best possible, because subsequent alterations are necessarily difficult and time-consuming and probably involve operational shutdown.

The lining's primary function is to support the ground and structures above, but it must also accommodate railway equipment such as cables and signals. Electrical equipment is sensitive to dampness, but moisture is almost unavoidable and the lining and fixings must be designed to minimise the effects. Even when the utmost precautions are taken to ensure watertightness in construction, slight ground movements may occur, resulting in seepage of water into the tunnel. An efficient drainage system should be provided, but there may remain the problem of atmospheric humidity and condensation of water on cold tunnel surfaces, more particularly in the early stages of commissioning before air movements and heating from traffic are effective. A fully operational ventilation system may therefore be an early requirement.

7.4.1 Tolerances

The size of the tunnel will usually be the smallest area within which the rolling stock and trackside or overhead equipment can be accommodated with appropriate clearances—further discussed later. In addition, tolerances for various errors of construction must be provided.

1. The first source of error may be in the setting out from the survey. No great errors should arise, but the transfer of lines and levels to the working face or to the shield is done in difficult conditions under pressure of time, and involves measurement of line, level, and shield attitude (see Volume 1).

2. A tunnel in soft ground, whether machine excavated or shield driven or without a shield, cannot be excavated and built to a perfect alignment and profile without intolerable delays. The more rapidly the excavation is advanced the more difficult it is to avoid departures from the specified line and level. If a shield or machine is found to be out of line it cannot be immediately adjusted to its correct position; it must be deflected so as to aim at coming on to line again a short distance ahead yet without overshooting into a reverse error.

Fig. 7.6 – Victoria Line. Brixton Station.

3. The lining introduces a third source of error, although with some opportunity for correction of minor errors of excavation particularly if relatively thick *in situ* concrete lining is used. Where precast segmental lining is to be used, errors may arise in the segments themselves from casting tolerances and also in erection to a true circle and in packing of joints. Imperfections in one ring add to the difficulties of erecting the succeeding ring correctly.

4. Following the lining of the tunnel, stresses in the surrounding ground readjust themselves and loading of the lining normally causes slight movement; the tunnel usually 'squats' by a small amount. Further movement is to be expected if another tunnel is built in the vicinity.

All these causes may accumulate to an error of several centimetres. The more experienced the tunnel miners and the surveyors and engineers, the more accurate the work is likely to be, but the more quickly the tunnel is advanced for economic reasons, sometimes for safety reasons, the more difficult it is to keep the deviations small.

In specifying a contractually acceptable tolerance all these factors must be taken into account, but even where the tolerance is exceeded it may be wise to adopt expedients which avoid rebuilding of the lining.

On completion of the lining a detailed survey is made from which a 'wriggle diagram' is produced (see Volume 1). A new alignment of the track centre line can then be calculated to utilise to the best advantage the clearances actually available, and to reduce or eliminate any necessity for rebuilding. The importance of experienced staff and their value to all parties will be apparent. Long experience on the part of all concerned allows the specified tolerance in London to be as little as 25 mm for hand excavation or 38 mm for mechanical excavation but with relaxations as suggested above.

The problem of construction tolerances has been described in the context of a bored tunnel in soft ground, but the same principles apply, although in lesser degree, for rock tunnels and to cut-and-cover; namely, allowances for setting out, excavation, structural lining, and ground movement. Operating tolerances are discussed below in terms of structure gauge and loading gauge.

7.4.2 Trackside equipment

Equipment within running tunnels should be kept to a minimum, because of the difficulties of access for maintenance and in the event of failure or accidents. As much as possible is therefore housed at stations. The necessary minimum in the tunnel, in addition to cables, is likely to include:

1. switchgear for isolation of traction feeds;
2. signal heads;
3. trainstops or similar devices for emergency braking of trains, and
4. disconnector boxes for track circuit and equipment connections.

The equipment must be designed to fit within the space available in the tunnel, leaving specified clearances which have to be accurately checked in relation to actual rail installation, allowing also for superelevation and curves horizontal and vertical. Design of covers for boxes housing equipment is of importance, to seal equipment against dust which is likely to have a high metallic content and to ensure that the covers cannot be left in a position encroaching on clearances.

7.4.2.1 Current collection systems

Metro systems, having relatively low route mileages, show a general preference for conductor rail systems. Overhead H.T. systems impose extra tunnelling costs because of necessary clearances.

If the conductor rail system is adopted there is a choice between 3-rail and 4-rail systems, the former being naturally the preference of engineers responsible for installation and maintenance costs of permanent way. For signal engineers there are advantages in the fourth rail system as this permits separation of the signalling currents in the running rail from the traction currents in the conductor rail; also, impedance bonds required in the third rail system can be eliminated. For close headway systems it may become more economical to provide a fourth rail.

The 4-rail system has a further benefit in permitting continuous monitoring of both poles of the traction system with earlier detection of faults and possible rectification before the service is affected. Recent development of audio frequency track circuits affect the balance of advantage.

7.4.2.2 Cable runs

Cable supporting systems usually consist of brackets, ducts or troughs. It is good practice to segregate low power signal cables and equipment from higher powered cables of the traction system by siting them on opposite sides of the tunnel. Fire precautions are important, particularly against the accumulation of dust round cables and in the adoption of cable insulation which minimises the risks of fumes, smoke and flame, while possessing electrical and mechanical strength. Thermal expansion and contraction is not a major problem in the relatively constant temperature of a tunnel.

Transition points between running tunnels and station tunnels require careful design of cable routes to preserve accessibility and avoid sharp bends. Segregation of signal cables from power cables should be maintained at these points, and also under platforms and in cable shafts.

The provision of suitable fixings for cable brackets or other supports must be considered in the design of the lining. Figure 7.2 shows the arrangement of brackets provided in the Victoria Line running tunnels. The frequent bolt holes of bolted cast iron linings offer a ready answer but even then it is necessary to provide adjustments which take account of the 'roll' of the iron and such local effects as superelevation which may cause the cables and equipment to encroach on clearances. Bolted concrete segments usually are less convenient in this respect, and with other types of lining, may make necessary the provision of special fixings for brackets, with scope for adjustment.

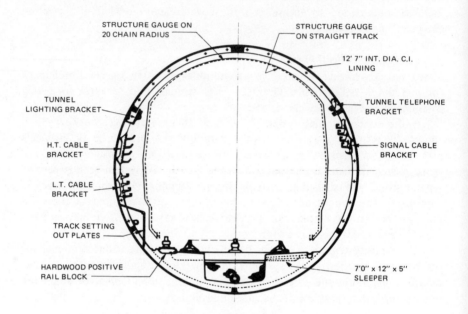

Fig. 7.7 – Victoria Line: cable brackets in tunnel.

7.4.2.3 Walkways

Emergency arrangements for safe discharge of passengers from a train immobilised in a tunnel between stations must be carefully planned. The emergency may be mechanical or electrical breakdown, or an obstruction in the tunnel, or fire. With decreased numbers of operating staff consequent on automatic operation in modern metros, the provision of easy means of escape becomes of increasing importance.

The ideal is an accessible platform on one side, at or near carriage floor height, but this is very expensive to provide both in tunnelling space and in its construction. While some metro systems are obliged by law to provide such walkways, in many there is no such platform and no space; the coaches are connected by a through corridor and passengers are expected to walk to the end of the train and there descend and walk along the track to the nearest station. For such circumstances the absence of a fourth rail is an advantage.

Where walkways are provided their design must provide for the accommodation of cables and pipes for the various services.

7.4.2.4 *Lighting*

It is not normally necessary to operation for running tunnels to be lighted, but reliable standby lighting for use in emergencies and also for use in night maintenance is required. The standard and quality of station lighting is of major importance in the contribution it makes to an attractive environment, over and above its narrowly utilitarian function.

7.4.2.5 *Communications*

An emergency telephone connection accessible between stations to drivers and guards of trains should be installed. It may be by wires carried on brackets fixed to the lining, or possibly by a suitably adapted radio system.

7.4.3 Noise and vibration

Reduction of noise experienced by passengers in tunnelled metros has been a problem actively investigated for many years. Noise originates in motors and in axles, wheels and interaction with the track. The hard surfaces of tunnel lining and track reflect sounds readily, unless treated with sound absorbing material (which should be non-toxic), and in the confined space of a tunnel the noise transmitted into the passenger coaches may be excessive.

Vehicle design can effect some improvement as can the use of rubber in the wheel mountings. Sound absorbent panels, acting also as baffles, may be fixed at wheel height as close to structure gauge as possible and the tunnel surfaces may be lined or sprayed with suitable material. Reduction of exposed area of concrete surfaces at track level and increase in ballast is also helpful. Careful sealing of openings in the body of the car together with careful design, construction and maintenance of track, wheels and suspension are even more important to reduce the noise experienced by passengers.

Large rolling stock allows greater opportunity of insulating the coaches from noise generated between wheels and rails. In the smaller London tube stock the tops of the bogies are below the passenger seats and little space is available for efficient sound insulation. Where stock is larger it is possible to design coaches without opening windows and to install air conditioning.

An underground railway may produce low frequency vibrations, or rumbles, which are occasionally heard or felt, particularly at night in buildings over the track. In Great Britain it has usually been found that this new traffic noise, initially considered obtrusive, is soon accepted as background noise which is rarely noticed.

The vibrations, transmitted through the ground, are not harmful to sound buildings. Nevertheless, there is a problem for which solutions are being sought. Rubber mountings for the tracks, plastic sleepers, heavy double sleepers and other devices to isolate the track from the ground are among the methods being tried. The extra expense is substantial and benefits are uncertain, more especially as it is only in special cases that objections are serious. Promoters of new metros are sometimes obliged to adopt such measures, but many cities maintain that the expense is not justified.

7.4.4 Drainage

Drainage is provided along the invert of the tunnel to dispose of all water from seepage or spillage. The gradient of tunnel, track and pipes should be at least 0.25%, leading to a sump and pump, which may be midway between stations. An important function of good drainage is to ensure reasonably dry conditions for track circuits, used to detect and register the presence of a train. Rail fixings to sleepers have to be designed with this in mind, particularly in damp areas. The vertical profile of a new line often makes it necessary to provide a drainage sump between stations. Such sumps should be pumped out automatically, but with appropriate warnings of pump failure. They will be discharged either to the nearest station sump, or direct to the surface, utilising a cable or vent shaft.

7.4.5 Track support

Conventional railway track support comprises cross sleepers of timber or concrete laid on stone ballast. This distributes loads and allows for efficient drainage. In the open it is easily installed and adjusted to the required line and level; regular maintenance is necessary and checking and readjustment. In tunnels these operations become much more difficult because working space and times of access are restricted and tolerances are small and therefore require more frequent checking and reinstatement; concrete support is therefore preferred.

The requirements to be satisfied as far as practicable by the track system adopted include:

1. Minimum tunnel size, and therefore minimum practicable clearances and tolerances.

2. Low cost of track installation.

3. Accurate installation requiring accurate and secure setting of rails before concreting in place.

4. Effective drainage, with particular reference to electrical controls.

5. Minimum operating maintenance with provision for vertical adjustment of at least 10 mm to allow new and worn rails to be matched up. Spring clip fastenings may be preferred to screwed fastenings in that the latter must be periodically tested for tightness.

6. Long term reliability.

7. Low noise level in coaches.

8. Minimal transmission of vibrations to ground.

9. Smooth track bed to prevent accumulation of dust which can ignite and smoulder.

Various designs of fastenings and concrete work are available, but the most suitable for any particular metro will depend on rolling stock, tunnel structure, skills and resources available, climate and other special factors. Whatever system is adopted involves some compromise between the differing requirements, and will be capable of progressive development in actual use.

Noise is greater with concreted than with ballasted track and increases significantly with speed. Below about 50 km/hr noise from a concreted track is not likely to be serious, but where speeds are higher, further measures become necessary as discussed above.

7.4.6 Structure gauge

The loading gauge is the profile beyond which no part of a train may project.

The structure gauge is the profile within which no part of a structure or fixed equipment may encroach.

It is the structure gauge that directly concerns the tunneller, but its profile is derived from that of the loading gauge by addition of safety clearances.

The loading gauge is set out in relation to the running rails as a base and is developed from the maximum car profile (including if necessary any projection caused by swinging car doors) by addition of allowances in width and height, for:

1. Track movement, tolerances and rail wear.
2. Train displacements in motion due to spring deflections, tolerances, defects and wheel wear.
3. On curves, the overhangs arising from the geometry of the coach in relation to the curved track.

Track movement to be allowed will depend on the form of track support adopted. Obviously, conventional ballasted track with cross sleepers is liable to a degree of movement much greater than when sleepers or other supports are encased in concrete and demands correspondingly greater tolerances.

Car displacement in motion may be both horizontal and vertical and has multiple causes: in the springing and suspension of the vehicle and between bogies and frame; in wearing down of original new surfaces of rail and tyre; and in variations of track gauge. These effects can only be estimated on the basis of experience and will be affected by the standards of maintenance imposed and the degree of wear allowed before renewal.

As a conventional rectangular coach travels round a curved track the bogies follow the curve, but the middle of the coach 'cuts the corner' and overhangs on the inside of the curve, while the ends of the coach overhang on the outside. Once the design of the rolling stock is settled, these overhangs are calculable. Their order of magnitude may be indicated by the example of an ordinary rectangular coach running on a 400 m radius curve for which the overhangs are about 50 mm.

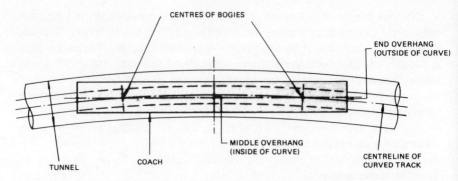

Fig. 7.8 – Coach displacement on curved track.

Minimum tunnel diameter, accommodating overhang, makes for economy of construction and can be approached by such measures in coach design as location of the bogies to equalise centre and end overhangs and rounding the ends and tops of coaches.

The structure gauge lies everywhere outside the loading gauge at minimum distances prescribed. Loading gauge and structure gauge for the Victoria Line are illustrated in Figure 7.9.

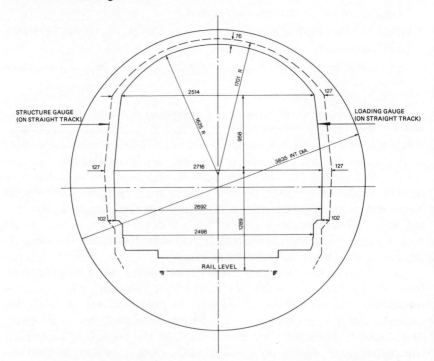

Fig. 7.9 – Structure gauge for Victoria Line. All dimensions are in millimetres.

Where the track is curved the allowances for overhang must be made, but in addition the structure gauge must be set out in relation to the rails which are normally superelevated or canted on a curve. The centre line of the tunnel structure and all critical points of its lining must then be offset by appropriate horizontal and vertical adjustments. The cant is built up gradually on the rails as a curve is entered and the tunnel centre line must correspondingly be progressively modified.

If the tunnel includes an emergency walkway it must, of course, be wholly outside the structure gauge but unless, exceptionally, it is to be used by staff when trains are running, it is unnecessary to provide the further clearances and safeguards which would be essential to give protection against passing trains.

7.4.7 Passenger flows and carrying capacity

A significant part of the cost of metros results from their having to carry peak passenger flows, which in most cities are limited to a few hours morning and evening five days a week. If they are to serve their purpose of reducing road traffic congestion all elements of the system must be designed accordingly: type of train, number of coaches, coach layout, length, frequency of service, together with station detail, length and width of platforms, widths of stairs and passages, numbers of escalators and lifts, ticket halls and turnstiles and ticket issue and collection. Some aspects of design have to be based on peak flows of as little as 15 minutes.

The limiting factor in train service is the duration of the station stop, which must be as short as possible if a close interval is to be maintained without the progressive delays resulting from trains held between stations. Numerous wide doors and clear space for free movement with cars are necessary for the quickest loading and unloading, but this means that seating capacity is limited and that coach designs are less suitable for suburban and longer journeys. Thus the highly specialised metro cannot satisfactorily be extended to distant outer areas. The practical carrying capacity of modern metro systems is between 30,000 and 50,000 persons per hour, or possibly more, in each direction, but light rail systems with a capacity of 15,000–20,000 per hour are now favoured in many cities.

Station design may be considered to start with the ticket hall, divided into the 'free area' outside the ticket inspection barriers, and the 'paid area'. Ticket control in and out may be either manual or automatic. Visual inspection may in practice be twice as quick as automatic gate control but is then likely to be less effective. A rate of about 25 passengers per automatic gate per minute is a fair average but may well be exceeded where most of the passengers are thoroughly familiar with the system. An escalator fully used will require about 5 gates handling 7,500 persons per hour. Between the top of the escalator and the gates at least 10 m should be provided for circulation to avoid dangerous congestion. Booking hall area and layout must also be designed to suit the ticket issuing and checking methods with careful study of pedestrian flows.

In passages a crowd density of 1.4 persons per sq m allows a free flow at about 3.8 km per hour which is 88 persons per metre width per minute. For fixed stairs the corresponding rates are 62 ascending and 69 descending, and escalators running at 44 m/min (2.5 km/hr) have a capacity of about 133 per min per escalator. At increased passenger densities the rate of flow in passages slows down and at a density of 4.3 person/sq m is reduced to a shuffle. The necessity is apparent for proper proportioning of passage and stair widths and circulating areas and avoidance of sharp bends and constrictions if a station is to handle peak flows satisfactorily.

7.4.8 Construction resources

The design of structures and methods of construction for a metro will be influenced by the availability of resources, both materials and human skills. The principal materials employed in any tunnelling are steel, cast iron, concrete and timber. Steel can provide temporary supports for trenches and tunnels, and permanent linings either as structural sections or as reinforcement. Cast-iron is the material which, in combination with the Greathead shield, made possible the development of soft ground tunnelling both in London and New York, and continues to provide a material of great merit for segmental linings, particularly in water-bearing ground. Spheroidal graphited iron is a form of cast iron having special structural advantages. Concrete has almost completely replaced masonry and brickwork for *in situ* linings in rock and in cut-and-cover tunnels, and is also preferred to cast-iron for segmental precast linings in many metros. Timber is not a material of permanent construction but is still most important for temporary support during excavation.

The principal choice to be made at the design stage is probably that of the material to be used for the permanent lining. The merits are more fully discussed in Volume 1 and briefly below. Availability of precast segments, iron or concrete, in the right quantities and sizes at the right times has to be ensured at the design stage. The initial use of concrete segments in place of cast iron in London in 1935 resulted from difficulties in securing adequate supplies of cast iron for the extension of the Central Line. The materials for concrete segments are usually available locally and a special casting yard can be set up, but a very high standard of workmanship is necessary if satisfactory segments are to be produced in quantity. Concrete segments are likely to cost substantially less than cast-iron, but for difficult tunnelling in water-bearing ground cast-iron may well have the advantage.

The choice between timber and steel for temporary support may be influenced by availability, but steel may be definitely specified at the design stage for some purposes, particularly where temporary supports are to be built in, and settlement may result if timber decays.

Tunnelling methods to be adopted must also be considered in relation to the availability of skilled men, familiar with particular techniques or capable of being trained. The degree of mechanisation to be adopted in construction will depend on many factors, but among the most important is the ability to assemble, operate and maintain the machines. Experts and skilled men will be necessary and it is virtually certain that some of them must be brought in from outside. It may be that the heavy capital outlay for mechanised work cannot be justified and that the better course will be to employ much larger numbers of less skilled men using simpler methods, which in any case are more adaptable to unforeseen ground conditions.

7.4.9 Tunnel lining

The lining for a tunnel must be selected at an early stage as a part of the whole metro system. Problems of ground support, methods of construction, performance in use and problems of maintenance must all be taken fully into account, in addition to cost of supply and installation.

Ground conditions may vary very greatly within short distances and although there might appear to be an economy in utilising cheaper linings where the ground is favourable, using more expensive linings, such as cast-iron, and S.G. iron only where difficult ground is encountered, there are very real disadvantages, even if the point of change is ascertainable in advance. A change involving alteration of the excavated area is not usually acceptable where a shield or excavating machine is in use, and even if the same excavated area is maintained the disruption of the system of progress can be very detrimental. With precast linings segments are manufactured well in advance and stocks are built up so that lining material is available in the tunnel at all times without delay. A decision to change cannot be taken quickly unless duplicate stocks of the alternative lining have already been built up. It is therefore generally advisable to use throughout a lining adequate for the worst conditions. Exceptionally, where a short length of particularly bad, or heavily loaded ground is foreseen, it may be of advantage to construct this length with special linings, but the converse modification of using lighter linings for a short length of favourable ground is most unlikely to effect any economy.

It is therefore best to design the lining for the worst sustained conditions in the length of tunnel under construction. This does not preclude the use of different linings on different sections of a line, as for example the use of cast-iron in an inner city area under buildings and of concrete further out where building loads are lighter or where a more favourable choice of ground is possible.

Watertightness is another factor to be considered in choice of lining. Cast-iron can more readily and effectively be made virtually watertight than can concrete, where some porosity is almost unavoidable and where cracks are likewise liable to occur and are very difficult to seal. Even in apparently impervious clay and with cast-iron lining some seepage of water is almost inevitable. Drainage must be provided to prevent accumulation of moisture; good ventilation is also helpful.

The requirement that there shall be an accurate relationship between track and trackside equipment must also be met, so that shape of lining and positioning of fixing holes must be precisely specified.

In a tunnel with *in situ* concrete lining all these considerations must be taken into account in the design of the shuttering. With prefabricated linings the choice between cast-iron, S.G. iron, concrete and steel depends on these and other factors, more fully discussed in Volume 1.

7.4.10 Ventilation and draught relief

There are two aspects of ventilation which have to be considered in the design of stations and running tunnels. The major air requirement is for cooling of the tunnels to keep within acceptable limits the build-up of heat arising from the motors and from braking. The ground provides a substantial heat sink, but in the long term the temperature inevitably rises until the ventilating air supply becomes warm enough to carry away surplus heat and reach a heat balance. The second aspect, and in immediate effects the more important, is passenger comfort. This requires an adequate supply of fresh air at an acceptable temperature and a limitation on air velocity in stations where the passage of trains can cause strong and unwelcome air movements.

The necessary air supply is that which, at the maximum acceptable rise of temperature, will remove the energy generated by the trains, averaged over a long period. The problems will be different in temperate or in cold climates where a substantial temperature rise is acceptable, and in hot climates where any increase may be unacceptable in passenger areas. This will influence the choice of points of supply and extraction of air. In the Victoria Line the air is normally drawn in as fresh air through the stations and extracted through shafts midway between stations, but the fans are reversible and can be used to draw in fresh air between stations. The piston effect of trains in tunnels can cause surges of air at stations. Reduction can be effected in part by draught relief shafts in the tunnels close to the stations. There may also be passages and entrances particularly subject to such draughts; much can be done in the design and if necessary special additional air passages can be provided. Seven metres per second air speed is considered to be an acceptable maximum for most areas, but a limit of 4.5 m/s is to be preferred if practicable, particularly for a flow of cold air.

Refrigeration of metros is a problem to be considered in hot climates. The fundamental difficulty is that waste heat is generated in the operation, and therefore plant requires further cooling from outside the system. If refrigeration is effected in the passenger coaches, there is added heat to be removed by the tunnel ventilation, while much of the effect may be lost when doors are opened at stations.

7.4.11 Environmental standards

The functional and aesthetic aspects of metro architecture cannot be separated. Apart from the trains themselves, most of the visual impact on users is in stations.

Construction problems and costs, particularly in tunnel, impose severe limitations on space both in plan and in height. Lighting, ventilation and noise have all to be considered.

In any but very shallow stations there will usually be two levels, with a booking hall at or immediately below ground level and deeper station tunnels, with escalators connecting them.

Ideally, the layout of a station should be such that passengers entering from the street progress directly past the ticket office to and down stairs or escalator and on to the platform, with a minimum of turns. The flow capacity of the access route should be almost uniform, except that the entrance gates at the top should have slightly less capacity than the escalators, and any exit gates or barriers should have a slightly greater capacity, so that the escalators themselves do not suffer dangerous congestion.

Platform length will be fixed by the length of the longest train plus the stopping margin; width should be sufficient to accommodate easily the maximum number of passengers entering between trains and also those detraining. Passengers entering and leaving must have room to move and sort themselves out. Emergency margins of space are also necessary for such circumstances as the emptying of a crowded train which has to be withdrawn from service, while incoming passengers continue to arrive on the platform. A reservoir area at the foot of escalators may be provided but closure of entrance barriers in the ticket hall should also be possible. Distinctive features and colours for different stations may be considered aesthetically pleasing, but have also operating advantages in enabling regular passengers to identify a station readily from a crowded train and thereby to alight without delay. This is of course additional to the provision of clear direction signs and station names which should be bold and simple and should be easily visible to passengers whether sitting or standing in the trains.

Another aspect of station design is the elimination of odd corners and recesses which can readily accumulate rubbish and cause a fire hazard. The same consideration applies also to the running tunnels between stations.

In deciding on the appropriate finishes for walls, floors and ceilings throughout a station, materials must be chosen which are available in suitable colours and are hard wearing and easily cleaned. Fire resistance and behaviour in a fire are of great importance. Where the station structure is below ground, and more especially below the water table, perfect waterproofing is almost impossible because of the inevitability of minor movements of ground and the ultimate porosity of even the best concrete. The very smallest penetration of ground water will, over a period, stain a decorative surface and deposit solids as it evaporates; it is therefore often advisable to make allowance for this by applying the decorative finish to a false lining with a small air gap behind it.

Lighting within a station should be of a high standard, but should not be higher than is adopted for the passenger coaches. Emergency lighting for stations to an adequate level should be automatically available in the event of failure of the main supply. Running tunnels need not be lighted ordinarily, but there should be emergency lighting provided for the evacuation of passengers from a stationary train.

Ventilation, to provide cooling, or heating, and draught relief have been discussed above.

Fig. 7.10 – Tyne & Wear Metro – Station Tunnel:
Secondary linings: Wall: Vitreous enamelled steel fixed to brickwork.
Platform: tiles on concrete beams.
Track bed: concrete.
Far wall: curved concrete slabs, advertising panels, corrugated 'Cellactite' to shed water.

7.4.12 Integrated stations

Apart from the detailed consideration of the best layout and environment within stations, the greatest shortcoming of many city stations is their separation from the activities which, in fact, they serve. In most cases they simply spew out and suck in passengers to and from busy and often narrow and crowded streets. The reasons are clear: minimal property acquisition and disturbance are required, and it is in any case likely that the constructing authority has no adequate powers to buy and develop, or that complex negotiations with adjoining owners will prove abortive. It is usually the cheapest and easiest solution simply to provide steps leading down to a ticket hall and attracting the passengers directly off the street. This may well be the best method in old cities close to historic buildings where disturbance should be minimal, but where an area is being redeveloped or in a new city or underdeveloped suburb, much better arrangements should be possible.

An urban station above an underground railway might well be served by escalators connecting to a covered shopping arcade at ground level and also serving a covered bus stop; other escalators might lead to an upper floor level of the shops and connect to pedestrian ways above traffic level with bridges across the streets. Other such connections should serve important buildings and offices and centres of entertainment. The benefits to the community should be substantial if costs could be properly apportioned.

Various metros have been able to incorporate exits through big department stores to the benefit of all. Montreal (perhaps encouraged by the climate) went further during their construction in the 1960's by incorporating some of its down-town stations in integrated developments covering several hectares; at the same time the different stations were made more attractive by varied architectural treatments separately commissioned. Some other new authorities in recent years have moved in the same direction—Melbourne, Helsinki, Hong Kong—but more success has been achieved in providing interchanges with other forms of transport than in the integration of station design and construction with other city development.

Fig. 7.11 — Montreal Metro — Station Tunnel.

It is probably generally true that metro promoting authorities lack the wide powers which would be necessary to use the opportunities occurring when a metro station is to be built in a city area ripe for reconstruction. It would certainly add to the complexities and delays of planning, but the problems are rather organisational and legal than engineering and architectural.

7.4.13 Passenger safety

One of the most important aspects of safety on underground railways is the avoidance of overcrowding, more particularly on platforms when there are service delays. When the designed capacity is likely to be exceeded, close supervision and early restriction on the entry of further passengers becomes essential, and it is important that the designed layout includes properly sited gates and barriers which can be shut as appropriate to the particular emergency.

In some stations a portion of the ticket hall below street level may also function as a public pedestrian subway; it should be possible to shut off access to the station proper while the subways still function normally. This consideration may also apply during non-traffic hours.

A vital element of station crowd control is obviously continuous monitoring and good communication between members of staff concerned. It was formerly necessary to have a large staff using an elaborate telephone system, but it is now possible to operate efficiently with a smaller staff using a closed circuit TV system combined with a public address system, and possibly remote control of escalators and barriers. This is an aspect of the central control system described earlier in this chapter.

7.4.14 Fire precautions

Any major fire on an underground railway is a serious matter, but they are very rare provided suitable materials and maintenance techniques are used. Those fires which do occur are mostly caused by accumulation of dust or inflammable rubbish, which is very vulnerable to ignition by cigarette ends blown about by draughts from moving trains. Such fires are not usually serious and can be extinguished by use of water or sand.

Very serious fires can result from overheating of equipment or cables due to an electrical fault, a major hazard particularly where plastic coverings can give rise to toxic fumes, although much work has been done to reduce this danger. Water cannot be used to extinguish fires in electrical equipment, while sand or other methods may be difficult to apply in some locations. Use of fans to remove smoke may be possible if the draught created does not intensify the fire.

With modern rolling stock fires are very rare and can usually be dealt with by use of portable extinguishers, or in electrical fires by cutting off the traction current.

It is a simple truth, not always appreciated, that the best precaution against fires is the avoidance of any materials which can burn, in the finishes and equipment of the metro and in the trains themselves.

Passenger emergency escape routes need to be fully considered in the design of the system, including contingency plans for exit to an adjacent station if an escalator link or other essential way out is blocked.

7.4.15 Flood protection
Underground railways, and particularly deep tunnels, are potentially vulnerable to flooding. In a city, long lengths are likely to be below the level of rivers, canals and sewers. The primary precaution must be to study carefully all records and to ensure that entrance levels at stations, and any openings for ventilation, cables or other purposes, are set above known and expected flood levels and are provided with barriers or other protection, to be used to raise the level if abnormal floods are experienced. Advance warning systems should be organised whereby information from all sources is directed without delay to a control centre from which instructions may be issued to stations for local action, or barriers may be closed by remote control. Drainage from the tunnel must be so designed that a hydraulic surcharge in the sewer to which it flows cannot cause a back flow. Flap valves by themselves do not give sufficiently reliable protection against major risks. In wartime, tunnels passing under waterways may be particularly at risk if the tunnel structure is breached. Floodgates may have to be installed at each end of vulnerable lengths to seal off the remainder of the system. Obviously electrical operation of gates would be provided but hand operation should also be possible to meet the contingency of power failure. At less vulnerable sites simpler forms of barrier might suffice.

7.5 CONSTRUCTION

Many of the problems of metro construction have already been discussed in considering planning and design. Tunnel construction methods have been described in earlier chapters, but some aspects special to metro tunnels may be examined in more detail. What follows is more particularly related to bored tunnels in soft ground.

7.5.1 Running tunnels
These comprise lengths of the order of 1 or 2 km between stations, usually for single tracks and therefore of minimum diameter, built to very small tolerances and requiring great accuracy of construction. Tunnelling may start from a shaft either in the vicinity of a station, or often midway between stations, in which case it may be advantageous to drive the tunnel right through the station site with subsequent enlargement. Shields or excavating machines have to be provided and assembled below ground in a shield chamber which may be formed as an enlargement of the future running tunnel. The special problems are those of

fitting the work conveniently into an overall programme, working very accurately in line and level, and handling excavated spoil and lining material on a restricted working site, possibly subject to strict regulation to minimise inconvenience to street traffic and to neighbours. Full mechanisation of tunnelling is only found advantageous for the longer drives and when ground conditions are reasonably uniform, because of the time consumed in setting up machines, their capital cost, their lack of flexibility in varying ground conditions, difficulties of accurate control of the geometry and the elaborate back-up facilities needed to handle peak output. A Greathead shield ('hand' shield) drive with precast segmental lining is likely to progress more slowly, but can be started more quickly and will utilise disposal facilities more uniformly over a longer working life. It is also easier to introduce the modifications necessary if ground conditions change unexpectedly.

7.5.1.1 *Crossover enlargements*
For operational reasons track crossovers are required at places where it may be desired to reverse a train, either in normal operation or in emergency. Their relevance to the opening of a line in stages has already been referred to. With single track running tunnels the enlargement necessary to include both tracks and the intervening space changes the scale of the tunnelling and presents special problems of excavation and support during excavation. It is likely to be best to drive and line the running tunnels straight through in the first instance and to use them as pilot tunnels to be enlarged to the full crossover size. For London standards the minimum diameter of the crossover tunnel is 9.6 m.

7.5.2 Stations
The term 'station tunnel' is here used to refer to a tunnel encircling rail track and platform and not to other tunnels forming part of the station complex, for passages, escalators, circulating areas, etc.

It is usually advantageous so to design station tunnels that the running tunnel shield or other equipment can be driven through the station length either before or after the station tunnel itself has been excavated.

Special problems arise in having to drive two or more large tunnels side by side and in providing cross connections between, and access to escalators. A common pattern is that illustrated in Fig. 7.12 where the station platforms are connected to an intermediate circulating area in tunnel which forms the lower end of an inclined escalator tunnel. Numerous access openings have to be constructed at the back of the platform through the tunnel wall.

The sequence of operations has to be planned with great care to fit together the timing of running tunnels and station tunnels with the excavation for escalators and circulating chambers, allowing proper time for the installation of escalators and other equipment in relation to the opening date for the line.

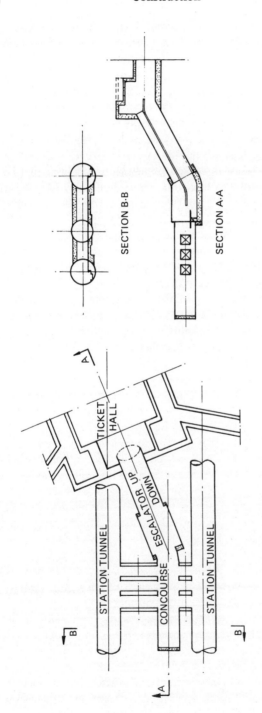

SECTION B·B

SECTION A·A

TICKET HALL

STATION TUNNEL

ESCALATOR UP

DOWN

CONCOURSE

STATION TUNNEL

PLAN VIEW OF STATION

Fig. 7.12 – Typical arrangement of tunnelled station with escalators.

It is not, of course, essential that the station tunnels for the two tracks should be side by side on the same level. There are instances where, in order to keep construction within the property boundaries defined by a street at the surface, station tunnels have been built one vertically above the other. Another important variation is at interchange stations between two lines which converge and are parallel through the station: the outward bound tracks from each line may be brought close together allowing simple cross platform interchange, and likewise the inward bound tracks. Many examples of this occur on the London Victoria Line interchanges shown in Fig. 7.13.

Where the station is not too deep the complexities of construction in tunnel may make it advantageous to build the whole station, or a large part of it, in open excavation from the surface. This method was used extensively in the Montreal metro, and made possible very open station layout, with lofty ceilings.

In tunnelled stations there are also examples of spacious construction. Moscow adopted a triple bay construction where normal station tunnels were opened out along the whole length at the back of each platform and supported by a line of columns and lintels carrying a connecting arch. In Leningrad running tunnels of normal diameter were similarly connected by a much larger arch spanning the platforms. In the London Central Line extension in 1935–40 an elaborate multiple arch station of this type was built at Gants Hill, providing a spacious station, but too costly for general use.

7.5.2.1 Escalators, concourses, passages

Construction of escalator tunnels can involve varied and interesting problems. The tunnel must be large to accommodate two or three escalators and operating machinery and is driven at a slope of 30° from the horizontal. For a pair of escalators the required diameter is about 6 m and for three escalators 7 m. The tunnel will almost invariably be driven downhill, and, although the lower part may be in the most favourable ground chosen for the running tunnels, the upper part may be in water-bearing alluvium where special expedients such as grout injection, or ground freezing, or construction in cofferdam may be necessary. A pilot tunnel may be employed not only to assist directly in the construction of the main escalator tunnel, but also possibly to provide earlier access to other work at the deep level.

Concourses, particularly at the lower end of escalators, may also require large diameter tunnels constructed as extensions of the sloping escalator tunnels and used to house the necessary escalator machinery.

Watertightness of these tunnels is particularly important to protect the installed equipment and structures, and also in the upper arch exposed to the public, where stains from even very minor seepage can be most unsightly and damaging to decorative finishes.

Cast-iron segmental lining has proved very satisfactory for such tunnels, all the joint faces being machined to accurate dimensions and in addition caulked.

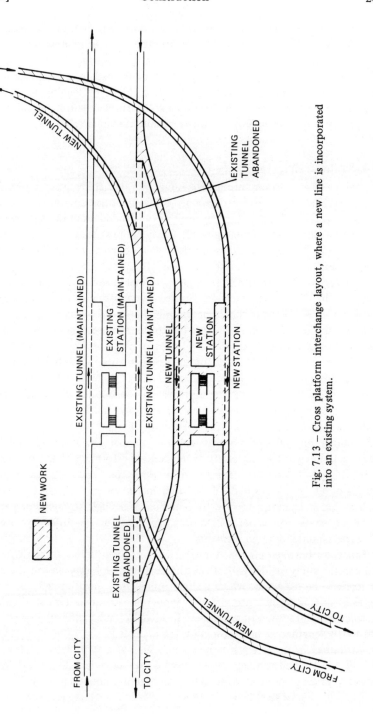

Fig. 7.13 — Cross platform interchange layout, where a new line is incorporated into an existing system.

The connection between escalator and concourse tunnels has often been constructed in specially designed segments forming a curve of small radius.

Passages also have to be tunnelled in stations at various levels, ranging from very short cross passages, connecting openings formed in adjacent tunnels, to long passages providing interchange between lines, or leading to exits at a distance. The passages may be rectangular or formed within circular tunnel linings; they may, if at shallow depth, be constructed in trench.

7.5.3 Openings and junctions

The making of an opening through the lining of a tunnel involves a redistribution of stresses. The ideal pattern of simple circumferential compressive stress in a circular cylinder is disrupted by an opening, and increases in the circumferential stress at the sides of the opening result. Wherever possible it is most desirable to make provision for such openings in the original design and to avoid any need for improvisation.

In a segmental cast-iron lined station tunnel openings for passages usually require to be 2½ to 3½ m wide, formed by omitting or removing segments from 4 or 5 rings. The rings should be so built that radial joints are in a horizontal line above the opening with adequate headroom for insertion of a lintel beam to redistribute the thrust. The sides, or jambs of the opening have to carry this extra thrust and may be formed of special segments, possibly of cast steel. A steel joist lintel is fixed and loaded by wedging, and a similar sill beam is required below to carry the corresponding up-thrust or reaction. It will usually be most practicable to build the tunnel in the ordinary way straight through the position of the opening, but making provision for later opening out by correct alignment of the joints, and also possibly by omitting one or two segments. Substantial propping is likely to be necessary throughout the forming of the opening, and careful filling with concrete, reinforced as appropriate and followed by pressure grouting of all voids excavated behind lintel and jambs.

Very small openings formed by omission of a single segment, or perhaps a pair of segments, for cable or pipe runs may often be simplified by dispensing with a special lintel or jamb segments.

Junctions between circular tunnels, whether longitudinal or at right angles, need careful study in design both in the complex geometry and in the stresses and loads to be expected. Where a small tunnel runs longitudinally into a large one, as a running tunnel into a station tunnel or a crossover tunnel, the simplest construction is provision of a concrete headwall of substantial thickness. Where a station is lengthened or a running track realigned or in other special circumstances, more elaborate work is necessary, such as a step plate junction where the large and small tunnel are connected by short lengths of intermediate diameters varying in progressive steps, each step embodying a small concrete headwall, or a heavy steel plate.

In some cases longitudinal junctions will be formed by use of taper rings of conical form either concentric or offset. These will be used where continuity of structure is particularly important due to heavy loadings and poor ground, and where waterproofing is very important as in escalator/concourse junctions.

For tunnels of different diameter joining at right angles an 'eye' may be provided in the larger tunnel within which the smaller tunnel is built. For this purpose the ratio of diameters should not be greater than 2 : 3 because the formation of such a junction between tunnels of the same size makes it necessary to open out a disproportionate area of ground, and the geometry of the connection has to be very accurate.

7.5.4 Construction programme

As with any other complex engineering project, if a metro is to be built efficiently and opened as soon as possible, a carefully prepared programme is necessary to ensure that each operation is started at the appropriate date and is so phased as to be completed in advance of subsequent dependent operations. This generalisation covers a very elaborate network of activities and interests. Legal powers must be obtained as soon as the line is decided, finance must be ensured to meet the anticipated expenditure, sites must be acquired, necessary materials and labour must be available and the work must be designed in detail so that contracts can be let and equipment ordered.

Key events and dependencies have to be identified, and times required for various activities estimated as closely as possible, so that a network incorporating those activities and the links between them may be established for use in forecasting and control. The general sequence of requirements for any section of running tunnel will include:

—acquisition of working sites
—detail design
—advance ordering of tunnel linings
—advance ordering of equipment
—letting of tunnel construction contracts
—completion of tunnel construction contracts
—installation of track
—installation of trackside equipment

Dates for these may be governed by priorities in relation to other units of work, in particular to station construction and also to access from adjacent lengths of running tunnel, and may also, where interchanges are involved, depend on modifications and diversions of existing tracks and tunnels.

Similar considerations are involved for station construction and depots, and for various aspects of equipment and also the supply of rolling stock. A very extensive and intricate network is built up such that computer control is likely to be necessary if it is to be fully utilised to monitor and regulate construction to the best advantage. Accurate reporting of departures from programme allows the network to be brought up to date; from this the ultimate consequences of delays can be seen and appropriate action can be taken by readjustment of the various activities.

Even without computerization flexible control of construction programmes is most valuable.

7.6 MAJOR SYSTEMS

The circumstances of every city vary so greatly that each produces a different metro system adapted to its own needs. When a system is initiated it is likely to be greatly influenced by existing public transport facilities; the first line may be an interconnection of main line or suburban subway termini, it may be the adaptation of tramways to a subsurface route, or a completely new system.

The factor precipitating a decision to construct a metro with underground lines is usually the congestion that develops in and about the centre of a city having a population of the order of one million people or more. It is, of course, difficult to make any precise comparison of city populations because traffic generating zones are likely to differ greatly from administrative areas, and in any case any metro line serves a narrow corridor rather than a city as a whole.

7.6.1 Route lengths

The major metro systems each having tunnelled routes in excess of 100 km are in cities with populations exceeding 6 million as follows: (The figures are from U.I.T.P. Handbook of 1975.)

City	No. of lines	Passengers per year x 10^6	Total	In tunnel	
New York	36	1266	407	232	Incl. P.A.T.H.
Paris	19	1225	248	178	Incl. Express
London	8	655	393	171	
Moscow	8	1628	150	132	
Tokyo	7	1498	150	125	Incl. TBTMG

It is further of interest to compare dimensions adopted for coaches in various systems and the tunnel sizes. The Glasgow system is included as showing the minimal size adopted in 1896. The limited height of the London tube stock, 2.88 m, is also of interest, in comparison with heights of 3.49 to 3.70 m adopted by many other cities and the intermediate height of 3.20 m in the new San Francisco (BART) system.

7.6.2 Dimensions of typical* coaches and tunnels

City	Length	Width	Height	Passenger Capacity Total	Seats	Tunnel
Glasgow	12.80	2.34	2.69	84	42	3.35 m dia.
Leningrad	19.70	2.70	3.70	170	40	5.10 m dia.
London (tube)	15.94	2.59	2.88	162	40	3.71 m dia.
Moscow	19.10	2.70	3.70	170	40	(7.6 m x 3.9 m
						(5.46 m dia.
New York	18.30	3.05	3.66	200	54	7.96 m x 3.89 m
Paris	15.15	2.40	3.49	157	24	7.10 m x 5.20 m
Tokyo	18.0	2.79	3.65	124	52	7.8 m x 3.9 m
Toronto	16.95	3.15	3.64	208	62	4.88 m dia.
Washington	22.86	3.05	3.30	175	81	

*Note that several of these cities have more than one type of coach in operation, and tunnels of various diameters.

BIBLIOGRAPHY

This chapter bibliography provides more especially references to British practice, for which very valuable and detailed descriptions of planning, design and construction problems are readily available. The meetings and publications of the International Union of Public Transport (U.I.T.P.) provide a world-wide forum, although concerned rather with operation than with construction. Further reference to the list of conferences in the main bibliography may be found helpful.

Baker, B., The Metropolitan and Metropolitan District railways.

Wolfe Barry, J., The City lines and extensions (Inner Circle completion) of the Metropolitan and District railways, *Min, Proc. Instn Civ. Engrs,* 1985, **81**.

Greathead, J. H., The City and South London railway; with some remarks upon subaqueous tunnelling by shield and compressed air, *ibid,* 1895, **123**.

Dalrymple-Hay, H. H., The Waterloo and City railway, *ibid,* 1899, **139**.

Biette, L., Underground railways: the metropolitan system of Paris;

Mott, B. and Hay, D., Underground railways in Great Britain.

Parsons, W. B., Underground railways in the United States, *Trans. Amer. Soc. Civ. Engrs,* 1905, **54**(F).

Parsons, W. B., The New York rapid transit subway, *Min. Proc. Instn Civ. Engrs,* 1908, **173**.

Gilbert, G. H. et al., *The Subways and tunnels of New York,* J. Wiley, 1912.

Jones, I. J. and Curry, G., Enlargement of the City and South London railway tunnels, *Min. Proc. Instn Civ. Engrs,* 1927, **224**.

The Post Office tube railway, London, *Engng*, 1928, **125** (Jan. 27, Feb. 10 and 24, Mar. 2 and 16).

Hall, H., The new Piccadilly Circus station, *Min. Proc. Instn Civ. Engrs*, 1929, **228**.

Groves, G. L., The Ilford Tube, *J. Instn Civ. Engrs*, 1946, **26** (Mar.).

The Stockholm Underground Railway: a technical description, 1958, A technical description of the Stockholm Underground Railway, 1964, Public Works Dept. and Passenger Transport Co., Stockholm.

Turner, F. S. P., Preliminary planning for a new tube railway across London, *Proc. Instn Civ. Engrs*, 1959, **12** (Jan); *see also* discussion, **13** (Aug.).

Jackson, A. A. and Croome, D. F., *Rails through the clay: a history of London's tube railways*, Allen & Unwin, 1962.

Howson, F. H., *World's underground railways*, Ian Allan, 1964.

Dunton, C. E. *et al.*, Victoria Line: experimentation, design, programming and early progress, *Proc. Instn Civ. Engrs*, 1965, **31** (May); *see also* discussion, 1966, **34** (July).

Havers, H. C. P., *Underground railways of the world*, Temple Press, 1966.

Follenfant, H. G., *Underground railway construction*, London Transport, 1968.

Follenfant, H. G. *et al.*, The Victoria Line, *Proc. Instn Civ. Engrs*, 1969, **Suppl.**; *see also* discussion, 1970, **Suppl.**

Howson, F. H., *The rapid transit railways of the world*, Allen & Unwin, 1971.

Nock, O. S., *Underground railways of the world*, A. & C. Black, 1973.

Follenfant, H. G., *Reconstructing London's underground*, London Transport, 2nd edn., 1975.

Associated Engineers, *Subway Environmental Design Handbook*, U.S. Department of Transportation, 1975.

Edwards, J. T., Planning and design of the Hong Kong Mass Transit Railway, *Proc. Instn Civ. Engrs*, 1976, **60** (Feb.); *see also* discussion (Aug.).

Jobling, D. G. and Lyons, A. C., Extension of the Piccadilly line from Hounslow West to Heathrow, *ibid*, 1976, **60** (May); *see also* discussion (Nov.).

Railway and tracked transit system noise, Proceedings of workshop, Derby, 1976, *J. Sound Vibr.*, 1977, **51** (3), Proceedings of 2nd workshop, Lyon, 1978, *ibid*, 1979, **66** (3).

Bayliss, D. A., *The Post Office railway London*, Turntable Pubns., 1978.

Rogers, L. H. (Ed.), *International statistical handbook of urban public transport*, International Union of Public Transport (UITP), Brussels, 1979.

Dasgupta, K. N. *et al.*, The Calcutta rapid transit system and the Park Street underground station, *Proc. Instn Civ. Engrs*, 1979, **66** (May); *see also* discussion, 1980, **68** (Feb.).

Cuthbert, E. W., The Jubilee Line – 1, the project.

Lyons, A. C., ditto – 2, Construction from Baker Street to Bond Street and from Admiralty Arch to Aldwych.

Bubbers, B. L., ditto – 3, Construction from Bond Street station to Admiralty Arch, *ibid*, 1979, **66** (Aug.); *see also* discussion, 1980, **68** (Feb.).

Edwards, J. T. *et al.*, Hong Kong mass transit railway modified initial system: system planning and multi-contract procedures.

McIntosh, D. F. *et al.*, ditto: design and construction of underground stations and cut-and-cover tunnels.

Haswell, C. K. *et al.*, ditto: design and construction of the driven tunnels and the immersed tube.

Taylor, R. L. *et al.*, ditto: design and construction of above-ground works and trackwork, *ibid*, 1980, **68** (Nov.).

Smyth, W. J. R. *et al.*, Tyne and Wear metro, Byker contract: planning, tunnels, stations and trackwork, *ibid*, 1980, **68** (Nov.); *see also* discussion, 1981, **70** (Aug.).

Howard, D. F. and Layfield, P., Tyne and Wear metro: concept.

Bartlett, J. V. *et al.*, ditto: management of the project.

Tough, S. G. *et al.*, ditto: design and construction of tunnelling works and underground stations.

Nisbet, R. F., ditto: interchanges and surface stations, *ibid*, 1981, **70** (Nov.).

Roscoe, R. B. *et al.*, Melbourne Underground Rail Loop, (11 papers), *Fourth Australian Tunneling Conference*, 1981.

8

Railway Tunnels

8.1 GENERAL CONSIDERATIONS

It is probably the railway tunnel through high ground that is thought of as typical when a tunnel is spoken of without other qualification. The Shorter Oxford dictionary in defining 'tunnel' uses the phrase 'now most commonly on a railway'. This obviously relates to public experience of tunnels in the 19th and early part of the 20th centuries, and to the days before road traffic developed to its present importance and resulted in the proliferation of highway tunnels. The associations of the word 'tunnel' may well become different in future.

From the first railway onwards tunnels have been an essential feature, primarily to provide a track with a limited gradient. In Volume 1 the early history of railway tunnelling has been described and some general characteristics of railway tunnels have been outlined. The fundamental reasons for tunnelling are topographical, with two distinct types

(1) to pass through rather than over a natural obstacle such as a hill, ridge, escarpment or mountain range, and

(2) to pass under a river, estuary or sea channel.

(1) Hill or mountain tunnels are the most frequent and characteristic type. Their function is to minimise the heights to be climbed and to keep gradients within acceptable limits, without involving lengthy deviations.

(2) Subaqueous tunnels, on the other hand, impose descents and unwanted gradients on the railway. On main lines (excluding metro systems) they are comparatively few in number, but are usually long, difficult and costly. They are therefore only constructed where the rail link is of major importance and alternatives are impracticable, or excessively long.

At the extreme for mountain tunnelling are the crossings of high mountain ranges as in the Alps, and elsewhere, and for subaqueous tunnelling sea crossings such as the Channel Tunnel and Seikan Tunnel described in Volume 1.

8.1.1 Urban tunnels

A third category which might be adopted is that of tunnels under urban areas. These are usually dictated by topography and as such need not be considered separately. Examples include the very earliest main line tunnels at the Liverpool terminal of the Liverpool-Manchester railway (1830), and the Hudson river tunnel in New York. In so far as other urban tunnels are not topographically determined, they resemble closely Metro tunnels in concept and construction. The principal differences may include:

—larger cross section
—terminal stations
—no intermediate tunnelled stations
—less frequent services
—longer and heavier trains
—non-electric traction
—use by goods trains.

New York City provides an interesting example of the dominance of topography. The main line from west to east has to pass in tunnel under the Hudson River to reach Manhattan Island and, after crossing the narrow island, to pass under the East River, again in tunnel. The Pennsylvania Station, near the heart of

Fig. 8.1(a) — The first railway tunnels, at Edge Hill, Liverpool (1830): Terminal tunnels, operated by rope haulage, comprising a passenger tunnel up to Cross Street and a goods tunnel descending to Wapping dock (see Vol. 1, Ch. 1).

Fig. 8.1(b) – Brunel's Thames Tunnel adopted for railway operation in 1869 (not the same Wapping as in Liverpool!).

the city, is therefore at a low level, 12 to 20 m below the street and is a through station between tunnel approaches rather than a terminus.

Melbourne's new Underground Rail Loop provides within the city centre better suburban services by eliminating the delays and congestion of reversing trains at termini. It also provides new stations giving better and more frequent access. Except in respect of the large tunnel cross section which is about 6 m diameter, the new Loop has all the characteristics of a Metro: primary use by passengers, frequent stations under city streets with easy access, electric traction.

The Paris express Metro also has the large cross section and wider spacing of stations more characteristic of main lines.

The generalisation about a smaller cross section in metros is only relevant where they are self contained systems whose use is restricted to their own special rolling stock and loading gauges. Main line tunnels must normally be capable of accepting all stock in use and must often allow clearance for hinged carriage doors swinging open and overhead clearance for electrical equipment or for the ventilation of diesel fumes or smoke. This larger and deeper cross section increases the tunnelling difficulties and costs, not only in respect of the excavation and lining, but by making more difficult the safeguarding of surface structures and by the necessity to go deeper below obstacles.

Fig. 8.2 – Potters Bar tunnels through London Clay. Old brickwork tunnel and modern duplicate employing expanding segmental concrete lining. (*British Rail*)

　　　Important terminal stations require so much space for numerous tracks and platforms with facilities for reversing and various shunting operations that tunnel construction is not practicable, but the lines radiating out from a city terminal are likely to be forced into tunnel if they run across the topographical grain of the country. New York City has already been mentioned. London also exemplifies this; the Great Western railway following the Thames valley could be constructed without any tunnelling near the city, but the North-South lines were immediately involved in tunnels. Stephenson's London–Birmingham railway out of Euston Station started northwards in deep cuttings and then had to be tunnelled under Primrose Hill and again further out at Watford. To the north east there had to be a sequence of three tunnels through London Clay at Potters Bar which were built as double track tunnels in 1849–50 and duplicated in 1959. The various lines southeastwards to the coast all had to pass through ridges, which required construction, among others, of the Penge tunnel through London Clay under Sydenham Hill in 1863, the Merstham tunnel through Chalk in 1841, the Bletchingly tunnel through Weald Clay in 1842. These are all within the general classification of hill tunnels and are not typically urban, because they do not involve the special problems of working in closely built-up areas and under streets and buildings.

　　　In urban conditions construction of main line tunnels as compared with Metros is simplified by the absence of frequent intermediate stations having problems of access and restrictions on gradients. The lower frequency of service and greater clearances make minor maintenance during traffic hours a possibility. On the other hand ventilation problems may arise from use of diesel power, or more rarely steam.

8.1.2 Hill tunnels

There is a wide spectrum from the relatively short and shallow tunnel which cuts through a hill spur or ridge in order to minimise gradients and shorten the route and avoid long or sharp curves, to the crossing of a mountain range with no practicable way over or round. Such tunnels will first be discussed in general terms, followed by a description of the special railway aspects determining design. Subaqueous tunnels will be described later.

　　　Just as it was the necessity for level lengths that compelled the canal builders to tunnel, particularly for summit levels, so the limitations of workable gradients forced the railways to do likewise. Where a ridge or escarpment is to be crossed the shortest tunnel is achieved by aligning it as squarely as possible at the narrowest width. There is some scope for keeping the tunnel short by climbing to the highest practicable level, but there must be a balance struck, taking into account both overall capital cost and operating cost and revenue earned. The shortest possible tunnel is not the best if it is attained only by unreasonable expenditure on long and deep cuttings, long and high embankments, steep gradients, and a climb to an excessive height.

Having fixed on the maximum acceptable gradient, the surface section of the line will be laid out to provide a close balance between the volume of material excavated from cuttings and tunnels and the volume required to build embankments, having regard to the factors for bulking and the suitability of the materials. Stability of embankments built with clay can be critical, and without the benefits of the science of soil mechanics, the early pioneers had many instances of continuing trouble in building high embankments, and also in stabilising cuttings.

In establishing such a line through hilly country the point of change from deep cutting to tunnel has to be decided with comparative cost as an important factor. In the early days, before development of major excavating machines, it was considered that the change should be made when cutting depth exceeded about 20-25 m, provided that the cutting could be relied on to stand at a slope of about 1½:1. Obviously, in unstable ground where a flatter slope becomes necessary, the volume of cutting increases considerably and a shallower depth justifies tunnelling. In rock which will stand up more nearly vertical, the excavated volume is correspondingly less at any depth, but tunnelling problems are different and a new balance of advantage must be sought.

The economics of excavation have changed somewhat with modern machine methods, but even where open cutting may appear relatively cheaper than tunnelling, the problems of drainage of deep cuttings, the stability of high side slopes, and the taking of a wide area of land are likely to be more decisive in the choice than the narrower considerations of construction cost.

8.1.3 Mountain tunnels

This balance between cutting and tunnelling costs is only one aspect of the much larger and more complex problem of the best route across a mountain range where there is no way round and a surface route would involve impossibly severe gradients and curves in addition to excessive exposure to snow and rock falls. A major problem is, again, at what level to drive the summit tunnel. A long deep tunnel offers the counterbalancing advantages of less height to climb at easier gradients and less severe exposure to the weather. The ultimate argument lies in the assessment of what capital cost can be financed by the expected traffic. The problem of connecting two established railway networks, as in the Alps, was very different from that of driving a new line such as the Canadian Pacific Railway through undeveloped country. The original Alpine tunnels were deep and long because that offered immediate operational savings, whereas the CPR across the Rocky Mountains used very steep gradients and sharp curves because traffic initially would not be of sufficient volume to effect large total savings. At Kicking Horse Pass, approaching the summit, the original line was at a gradient as steep as 1 in 25; this was later reduced to about 1 in 45 by looping the line in two spiral tunnels driven into the mountainside when traffic became adequate to justify the cost, even though these loops added 4 miles to the length of line. Similarly, on the original transcontinental lines of the USA, zig-zag layouts involving reversals were adopted; they were costly to operate and slow but saved heavy capital expenditure.

Fig. 8.3 – Swiss railway tunnel in approach to Lötschberg and Simplon Tunnels. Originally constructed for single track, but with provision as shown for widening to double track. (*Berne-Lötschberg – Simplon Railway.*)

As the systems developed and traffic grew, a number of long summit tunnels were built to improve operation. In the USA these include the Moffat tunnel (1928) 10 km long near Denver in Colorado which replaced a line at 3500 m over Rollins Pass, and the Cascade tunnel (1929) 12½ km long near Seattle in Washington replacing a shorter tunnel and reducing the summit level by 150 m.

The same pattern of long railway tunnels being constructed, as later improvements in mountain areas to reduce steep gradients and improve operationally expensive routes, has been followed also in New Zealand and in South Africa.

The original Trans-Iranian Railway from the Persian Gulf in the south to the Caspian Sea in the north via Tehran was largely built in the years prior to 1938 and required much tunnelling, particularly in the Bakhtiari mountains in the south and the Elburz mountains in the north including a number of figures-of-eight with spiral tunnels and with gradients up to 2.8%.

The line northwestwards from Tehran to Tabriz, a distance of about 700 km, was built as a single track with curves as sharp as 250 m radius.

It was to be largely rebuilt to modern standards for high speed traffic with double track minimum curves of 5000 m to 1350 m and with gradient limited to 1.0%, except in the mountainous area approaching Tabriz where gradients of 1.4% are allowed. Inside the mountain tunnel portals a gradient of at least 0.8% for the first 250 m is sought to ensure rapid flow of drainage water and minimise ice formation. The distance would be reduced by about 100 km by a new route through the mountains from Poledokhtar to Tabriz.

In Japan major railway developments have been in progress since 1960. A policy of linking all parts of the four main islands by modern high speed trains has been pursued very actively. Because the country has many mountainous areas bordered by narrow and densely populated coastal strips, tunnelling on a large scale has been necessary to ensure both the easy gradients and large radius curves demanded for high operating speeds. Initially, for the Tokaido line from Tokyo to Osaka the design speed was 210 km/hr and the ruling gradient 2% and curve 2500 m. On its extension first to Okayama and then through the Shin Kanmon tunnel to Hakata on the island of Kyushu, design speed has been increased to 260 km/hr and construction limits have been progressively modified to 1.5% gradient and 4000 m radius (but with an exceptional 1.8% and 3500 m in the Kanmon tunnel). Another line runs NW from Tokyo to Niigata. The proportion of tunnel to open line is extremely high in these lines and the tunnels are among the world's longest.

8.1.3.1 *Portal Cuttings*

In contrast to the lowland railway requiring a tunnel through a hill or escarpment which tends to approach the slope squarely, the high mountain railway, which makes use of a valley is to a great extent laid out on sidelong slopes. Deep cuttings in such ground are not usually practicable so that the problem of siting the tunnel portal is different. The ruling gradient determines the possible alignment up the side of a valley, cutting obliquely across the contour lines until they become too steep or too sharply curved. Sometimes a viaduct will be required, as across a side valley but eventually tunnelling becomes essential, initially perhaps only for a short length through a spur but eventually into the main mass.

These simple geometrical requirements of the topography are only the first step in laying out the centre line because the width of the formation is very significant on steep sidelong slopes, and is likely to require the building up of a bank on the low side while a notch is cut into the slope on the other side. The

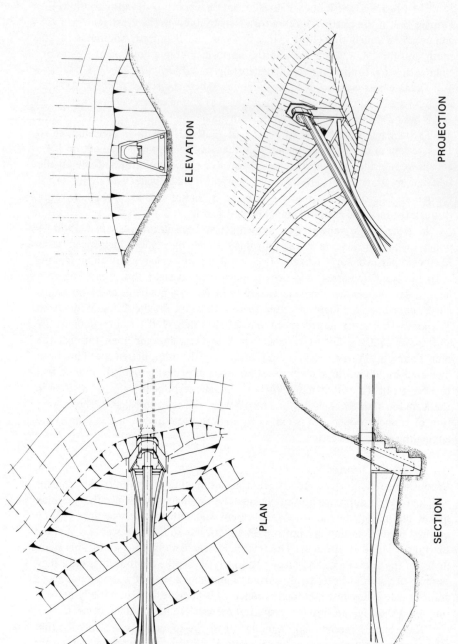

ELEVATION

PROJECTION

PLAN

SECTION

Fig. 8.4 – Portal and tunnel approached by bridge, proposed for Iran railways. Note transition problems at portal.

Fig. 8.5 – Avalanche protection on Austrian railway. *(Austrian Federal Railways)*

characteristics of the rock and the whole geology of the area together with its
exposure to rock slides, avalanches, snow and other conditions are of vital
importance. Another consideration in seismic areas is that high viaducts become
extremely costly to safeguard, turning the economic balance in favour of the
tunnel.

Siting and construction of tunnel portals also presents difficult problems
in the angle of approach and the penetration of loose and weathered rock before
reaching competent rock. There may be a need to protect the approach to the
portal against rock falls or snow by an extension of the tunnel. The operating
length of the tunnel will then exceed the bored length.

Such problems of siting are not peculiar to railway tunnels but apply also
to highways, which are, however, much more flexible in that steeper gradients
and sharper curves are acceptable.

8.2 ROUTE SELECTION

The promotion of a railway link through country where tunnelling may be
required will first be determined in relation to its potential usefulness commerci-
ally and socially, and also sometimes in the context of strategy or even in terms
of national prestige. The comparative merits of railway and highway must be
weighed and the whole project studied in terms of its capital cost, operating
cost, and benefits whether measured in terms of revenue or otherwise assessed.

Although the proposed railway may be seen primarily as a transport link
between its termini, which may be junctions with other railways, it will almost
certainly be required to serve also intermediate places. These requirements
are constraints on the choice of route, as are the various topographic features,
but they also provide opportunities for attracting profitable traffic and may
attract sufficient additional traffic to justify expenditure on higher standards
of construction and equipment.

Of first importance in selecting a route between points to be served is the
adoption of the most direct alignment with minimum summit height, gradient
not exceeding a selected maximum, and curves within a prescribed radius.
Tunnels and viaducts may serve all these purposes by their independence of the
surface topography. Most lines require both, but the proportions as part of the
whole line vary greatly according to the ruggedness of the country, and the
extent to which the line accommodates itself to the grain of the country or cuts
across it. The whole economy of a line in construction and operation is dependent
on making the best use of the ground, including the adoption of tunnels.

Where a tunnel is adopted its siting and alignment to the best advantage of
the line as a whole may be a very complex problem. Apart from the choice
between a shorter tunnel at a higher level and a longer tunnel lower down on a
simple basis of construction costs compared with operational savings, the geology
of the area may influence very strongly the preferred tunnel line and level. There

may also be questions of intermediate access for construction or for ventilation or other operational reasons.

8.2.1 Topographical survey

The first requirement for planning a railway route and its tunnels is a topographical survey of as wide an area as possible. Adequate maps for initial needs may exist but progressively more detailed maps are required as planning and design proceed and the alignment is more precisely defined. The subject has been dealt with in general terms in Volume 1 but there are special aspects relevant to railway tunnels. At no stage can the scope of the survey be limited to the tunnelled length, but must be adequate to include long open approaches and potential areas of portal construction. In rugged mountainous country every detail of the alignment has to be studied as part of the whole. Viaduct, sidelong cutting, portal and tunnel interact and cannot be seen in isolation. Possible sites for vent shafts or access shafts may be important, also any watercourses or springs or standing water or drainage features. As for all tunnels, any buildings in the vicinity need to be located exactly.

For the preliminary stage of establishing technical feasibility and identifying possible alternative railway corridors, maps to a scale of 1:50,000 are suitable. Areas requiring more detailed mapping can be identified, and limitations imposed by gradients and curves can be examined.

For the next stage maps of the selected corridors at a scale of 1:10,000 allow feasible alignments to be confirmed, and unsuitable corridors to be eliminated, gradients and curves again being checked. Areas for more detailed mapping can again be determined, and the location and scope of major structures required can be identified.

Maps to a scale of 1:2000 allow the best alignment to be finally fixed and its geometry to be specified. The major structures can then be defined and quantities prepared for structures and earthworks.

For subaqueous tunnels similar procedures are necessary but there are additional aspects of prime importance which include
—accurate survey of bed levels
—determination of nature and stability of bed
—possibilities of deepening by scour or by dredging
—flood levels past, present and prospective
—tide and surge levels, normal and abnormal
—presence of waterbearing fissures and fault zones

8.2.2 Geological survey

The geological mapping of the route is just as important as the topography, in some cases more important. The same process of progressively more detailed study of narrower areas as for the topographic survey is desirable, but may not be practicable.

A general study of available information covering the possible routes, supplemented by a visit to the area, will help in establishing the preferred corridors but much will depend on the detail and accuracy with which general mapping has been done, and also on the rock exposures available for inspection or the extent to which they are masked by superficial deposits.

At the 1:10000 mapping stage, field reconnaissance of the alternative corridors is essential to assess the character and structure of the rocks and to prepare proposals for a detailed exploration of the preferred route with the aid of boreholes and geophysical survey.

As the 1:2000 mapping develops the need for further boreholes and tests, particularly at the sites of major structures will become apparent and should be organised, preferably as an extension of the main borehole programme.

Apart from the aspects of rock nature and structure essential to all tunnelling, as dealt with in Chapter 4, special railway problems arise, for example, at the tunnel invert, where squeezing clay may be a long term hazard unless an invert arch is provided, and where good drainage can be vital to a sound road bed.

Ground water can be detrimental to the electric transmission system unless proper precautions are taken, and also can reduce rail friction seriously, especially in conjunction with clay or if frozen, and thus reduce the haulage capacity of locomotives particularly on a gradient. In some climatic conditions dripping water can freeze into dangerous icicles and can coat the rails with ice.

The geological conditions at and near the portals are likely to require special study where there is danger of rock slides and of the presence of badly weathered rock.

Subaqueous tunnels demand even more detailed investigation of the possible effects of ground water both in the construction phase and in subsequent operation. Porosity of the strata and the presence of fissures can be critical to the feasibility of tunnelling methods and may influence the choice of lining to minimise seepage and consequent pumping installations.

8.3 GEOMETRY OF ALIGNMENT

The topography and geology of the area, together with commercial and industrial development and potential, constitute the external constraints and influences on the alignment. To the extent that they are known or assumed the geometry of the line can be fixed. The principal geometrical elements are:

(1) gradient
(2) curves
(3) tunnel cross section.

8.3.1 Gradient

The ruling gradient is of fundamental importance to the planning of a line and should be fixed provisionally as early as possible, even though it may later be found advantageous to modify it. It may be fixed to be the same as that on existing lines to which the new line will be linked and thus to suit the locomotives currently in use or available, but it is possible that overall economy will be achieved by the use of new and improved locomotives which can haul full train loads up a steeper gradient.

Gradients have commonly been described as: 1 in x, or as a percentage, but there is a growing practice of describing them as per thousand, using the symbol $^o/_{oo}$. A gradient of 1 in 100 may thus be displayed as 1.0% or as $10^o/_{oo}$. Unfortunately this last form, which has many merits, is not sufficiently distinctive, and is all too easily reproduced as per cent. The percentage is therefore normally used here. There is no well defined limit for gradient, but rail adhesion in relation to loco weight and power imposes practical limits. A railway as a whole is planned with a ruling gradient as a normal maximum although exceptions are possible where a critical length can be overcome by 'double heading' the heaviest trains. Nevertheless, gradients exceeding about 1% are not adopted without good reason, although much steeper gradients over 2% were used in the early Sarnia tunnel, and the Appenine Pracchia tunnel gradient is 2.5%. The 12½ km Cascade Tunnel has a gradient of 1.6% which causes difficult ventilation problems for diesel locos on the ascent; in order to ensure fresh air for the loco drivers a portcullis at the upper portal is shut and a fan is used to provide a flow of fresh air against the train until it is about to emerge. 1.8% has been referred to above for the Kanmon tunnel.

The factors ultimately controlling gradient are loco power and rail adhesion. The power required is proportional to the product of weight of train, speed and gradient with the addition of frictional resistance and wind resistance. This power must be provided in the locomotive and transmitted through its driving wheels to the rail, where the adhesion available is proportional to the effective weight carried on the driving wheels and to the coefficient of friction.

A tunnel may give rise to two particularly unfavourable factors, namely, low rail adhesion because of persistently damp conditions, and high air resistance to forward speed because of the restricted space. The consequence in such tunnels is that either loco power and weight must be provided in excess of what is needed on the open line, or tunnel gradients must be limited to, perhaps, two thirds of the ruling gradient elsewhere. Icing on rails has also to be considered, but that hazard is probably less in a tunnel than in the open, except for dripping water near the portal.

At the other end of the scale gradients must be adequate to ensure drainage, for which about 0.1% is a minimum. Except in subaqueous tunnels, where pumping of drainage is unavoidable, drainage by gravity should be ensured. Something steeper than the 0.1% minimum is to be preferred if closed drains are to be self cleansing.

Fig. 8.6 – Removal of ice from a tunnel in Canadian Rockies. (*Canadian Pacific Railway*)

8.3.2 Curves

Minimum curve radius on a main line railway is commonly fixed so that the cant at maximum design speed does not exceed some arbitrary limit. The amount of cant (C) for exact balance between the inward tilt and centripetal acceleration of a coach is given by the equation.

$$\frac{C}{G} = \frac{V^2}{Rg}$$

where G is rail gauge, V is train speed, R is radius of curve and g is gravitational acceleration. The minimum radius is likely to be some hundreds of metres, up to perhaps 4000 or more for high speed and the maximum cant may be as little as 50 mm but is unlikely to exceed 150 mm. The structure gauge may have to be rotated in the same degree.

This curve limitation has little direct influence on tunnelling. The small consequent increase in width of structure gauge makes necessary a corresponding increase in excavated width, and the setting out and survey are slightly more complex than for a straight tunnel. In difficult country, however, a sharp curve in the tunnel approach may be very helpful in permitting the tunnel to be aligned more favourably and its portals to be sited in better ground.

8.4 FORM OF TUNNEL

For the railway tunnel, there is a choice to be made between one double track tunnel and twin single track tunnels. There may, of course, be cases where the traffic foreseen does not justify more than a single track, possibly with provision for later duplication.

The choice will depend on tunnelling considerations and costs and also on operational requirements. Among the factors to be considered are:
(1) Construction
    (a) Total excavated volume
    (b) Length of tunnel
    (c) Stress interactions of twin tunnel
    (d) Separation of approaches and problems of alignment
    (e) Sequence of construction and problems of access, ventilation and drainage.
(2) Operation
    (a) Aerodynamics and ventilation
    (b) Safety — collisions and fires

The difference in width is that a double track tunnel requires a minimum width of about 8.5 m and a single track about 5 m after allowing for safe clearances. The minimum height over coaches is obviously the same but the difference in maximum height at the crown depends on the curvature of the roof arch and therefore on the characteristics of the rock and the lining adopted. With the normal horseshoe section the area of the double track tunnel is likely to be rather less than twice that of the single track tunnel, but, with a full circular section, which is more typical of subaqueous tunnels, the proportionate increase in height makes the area of the larger tunnel more than double. Cross sectional area is by no means the only measure of cost because excavation and support problems in tunnelling can increase disproportionately with size as discussed in Volume 1.

The length to be tunnelled is obviously very relevant to the decision. Twin tunnels require to be separated, and if the tunnelled length is short the extra cost of spreading the approach tracks is unlikely to be justified by any saving in tunnelling costs. The special circumstances in Metro tunnelling may be compared: the tunnels are long and separation of tracks may assist station design especially at interchanges.

Duplication of a single tunnel by a parallel tunnel involves tunnelling through ground in which the stress pattern has already been distorted and, in turn, causes a change in stress on the tunnel already built. The wider the separation, the less severe are the effects, and in this respect the widest practicable separation is obviously to be sought; one diameter of tunnel may be considered as a minimum in moderately good ground. It may be necessary for the tunnels to converge at the portal if the two tracks are to be accommodated on the same formation outside, but this can make more difficult the portal construction problems in what may be weathered and disturbed surface strata. In a long tunnel, of course, it is relatively easy to effect, in the tunnelling, as wide a separation as thought necessary as distance from the portal increases.

If twin tunnels cannot be made to converge at the portal, the tracks outside must be similarly separated, and, if in cutting, may necessitate separate deep cuttings eventually converging. If on a sidelong slope the levels of the tracks may have to be different. These geometrical problems of the alignment are of most significance in rugged country and for long tunnels.

In construction of the tunnel, there are benefits to be gained from separation into twin single track tunnels. One heading will almost invariably be driven ahead of the other, and the information on the behaviour of the ground should be of great benefit. If treatment of the ground, such as grouting or rock bolting, is necessary, continuity of specialist operations can be maintained by alternating between the two headings.

If the two are interconnected, ventilating air can be circulated up one tunnel and down the other, a particular advantage if fumes from explosives, or gases such as methane, have to be dispersed. Drainage also can be more easily maintained. In the event of falls of rock occurring behind the face access is much better.

8.4.1 Operation
Among the operational problems in long tunnels is the 'piston effect', the wave of compression ahead of a train, which is most significant in a single track tunnel

in which the ratio of the cross sectional area of the train to that of the tunnel is high. This is of increasing importance when operating speed is high, both in its resistance to the train and in the possible unpleasant effects to passengers from abrupt changes of pressure. It is more fully discussed in Chapter 9.

Planning of ventilation is of even greater importance where fuel burning locos are in use rather than electric traction. In single track one way tunnels of moderate length, the draught produced by train movement will usually ensure reasonable conditions by effecting a progressive change of air unless there is a persistently unfavourable wind. In a double track tunnel, on the other hand, the train movements will cancel out and change of air will be dependent on wind effects or forced ventilation, in the absence of which there is likely to be a stagnant area near the middle of a long tunnel.

Operational safety may also influence the choice. Head on collisions are obviously not to be feared in a one way tunnel; derailments on a single track are limited to the train concerned whereas with double track, both lines may be blocked. Fire likewise appears a lesser hazard in twin tunnels. There is the further advantage in any emergency of being able to use the second tunnel for evacuation of passengers and for rescue equipment, provided that there are cross connections available. Most maintenance and repair operations can also be carried out more efficiently with sole possession of a single tunnel, when traffic can be diverted to its twin.

8.4.2 Cross section

The size of tunnel to be excavated is based on the structure gauge for the line, more fully discussed in the previous chapter. It is a profile, set out in relation to the rails, within which no part of any structure or fixed equipment may encroach and is derived from the loading gauge by making appropriate allowances for such things as sway arising from springing of coaches, working tolerances and wear in suspension, wheels and rails, and then adding appropriate clearances at various points of the cross section. Where a walkway is to be provided this in turn adds to the area of the structure gauge. On curves, as explained in Chapter 7, a wider structure gauge is specified, but in tunnels designed for high speed running the curves are of such large radius that this allowance will be small. On curves also where the rails are canted, the structure gauge must be similarly canted. Another concept sometimes found useful is the 'clearance diagram' for tunnel excavation and lining, leaving room within for such fixtures as cable brackets and signals etc. and for varying cant on the track.

Area to be excavated

The excavated area will extend outside the structure gauge. Allowances must be made for:

(1) Specified construction tolerances
(2) Space for cables, ducts and brackets

(3) Overhead line installations
(4) Tunnel lining
(5) Track support
(6) Overbreak
(7) Other requirements.

1. Construction tolerances
The tolerance applicable to the finished lining may be as much as 150 mm for a high speed line because very little adjustment in wriggle diagrams can be accepted, and correction of the lining is likely to be difficult and time consuming and may be deleterious to its strength and integrity.

2. Cables, ducts and brackets
Cables will usually be accommodated in ducts or on brackets on the walls of the tunnel, preferably on both sides of the track, separating power cables from signalling and control cables.

3. Overhead line installations
The space necessary for high voltage (25000 V) wires and their supporting structures determines the crown height of the tunnel. Accurate and secure fixing in this difficult position must be ensured and also as far as practicable freedom from moisture. Special waterproofing work may be necessary in this area, more especially if winter conditions are such that icicles can form.

4. Tunnel lining
The thickness of the specified lining must be added to the excavation profile. Greater thickness may be required where bad ground is encountered. Depending on the ground, lining may or may not be required in the invert.

5. Track
The choice of track bed, discussed below, will determine the allowance to be made in the invert.

6. Overbreak
In all but the most precisely trimmed machine excavated tunnels there is bound to be some overbreak. It may be substantial locally in fissured rock and must be taken into account in the design of the lining and in calculating quantities of spoil and backfilling.

7. Other requirements
It is sometimes considered worth while to allow outside the normal structure gauge a further space to accommodate scaffolding for remedial work to the arch.

In seismic zones, specially large clearances may be specified at known fault breaks to simplify the correction of misalignment resulting from earthquake shocks.

Aerodynamic effects in high speed running may, in exceptional cases, justify a larger tunnel area. Refuges, at regular intervals may be specified outside the general cross section. The cross section finally adopted will have a smooth profile enclosing all the relevant allowances.

It will almost invariably have an arched crown which may be continued as a horseshoe shape down to track bed level. Straight vertical sides up to arch springing may be adopted. The invert construction will depend on the expectation of ground loading in the long term. Where there is lateral pressure the side walls must be strutted apart at the base and where uplift may develop as in clay, an arched invert lining must be allowed for. The ultimate shape for all round pressure is the circle, extensively used in subaqueous tunnels. The use of full face tunnelling machines also may impose a circular shape.

8.5 TRACK

The merits of ballasted track with sleepers on main line railways are such that other forms of track are only used in special situations.

In very many tunnels ballasted track is used, but some disadvantages arise and alternatives must be considered. The depth required for ballasted track is about 300 mm compared with 200 mm for a concrete slab. The tendency of ballast to move slightly under the impact of train loads may result in enchroachment on the clearances and make necessary more frequent track adjustment and repacking. This is a more difficult and potentially dangerous operation in the limited space of a tunnel, particularly if mechanical equipment is used, even if specially designed to work within the structure gauge. Renewal of track is likely to be necessary at more frequent intervals of time in tunnels because of the generally humid and perhaps corrosive atmosphere, and after renewal the necessarily precise setting and re-adjustment of the track is troublesome for the bedding-in period.

For a short tunnel, it is not desirable to introduce the discontinuities of a altered type of track, but in a long tunnel a continuous concrete bed laid to precise dimensions with modern slip-form machines has many merits. The track is retained securely in its required alignment, and requires a minimum of maintenance work. The rails may be secured to the concrete by resin grouted bolts if workmanship of a high standard can be ensured, and may rest on resilient pads to cushion impact forces.

A transition 'ladder', between ballasted track outside the portal and concrete slab track inside, can with advantage be provided. Good track drainage is necessary in all cases, preferably gravity drainage provided by the gradient of the tunnel.

Fig. 8.7 – Solid concrete bed for track in 5 km Heitersberg rail tunnel: Zurich–Berne. (*Swiss Federal Railways*)

8.6 SUBAQUEOUS TUNNELS

On main line railways subaqueous tunnels are few in number. In Great Britain, out of over a thousand British Rail tunnels only the Severn and Mersey tunnels, both opened in 1886, are subaqueous.

Tunnelling is not favoured for a river crossing by a main line railway if a bridge provides a feasible alternative. The fundamental reason, apart from difficulties of tunnelling, is that, to pass under a river, the railway must descend far below ground level and ascend again. This makes severe demands on locomotive power especially when a heavy train is stopped and has to restart against the gradient. Also, capital and maintenance costs are high.

The necessary descent of the tunnel is compounded of:

(1) Initial height of railway above water level

(2) Depth of water

(3) Added depth for scour or future dredging

(4) Cover over tunnel

(5) Height of tunnel from rail level to crown

(6) Depth of invert construction below rail.

These are likely to add up to at least 20 m. If a gradient of 1% is adopted the lengths of approach descent and ascent will be 2000 m each under level ground. At least half of this is likely to be in tunnel, making the tunnelled length equal to the channel width plus 2000 m.

The profile makes it necessary for all drainage to be pumped and the volume is likely to be substantial. Forced ventilation may also be necessary; the Mersey tunnel was an early example, in the age of steam.

These disadvantages do not apply in anything like the same degree to a Metro system because the general rail level is already well below ground level so that any descent will be very much less, or entirely absent. Also, Metros do not have the gradient problems imposed by heavy goods trains.

It is in estuarine conditions, with a broad alluvial plain, that tunnels are advantageous because bridges become difficult and costly to build. Shipping requirements demand long spans and high clearances, and foundation problems may be severe. It is no coincidence that where railway tunnels have been built under the Hudson river and East river in New York and the Severn and Mersey in England, all are estuarine.

For the very long crossings of straits by railways, tunnelling becomes indispensible. The Channel tunnel between England and France is an outstanding example of the concept, under discussion for over a century as a practicable project, and initiated in earnest on at least two occasions. The Seikan tunnel, to connect the Japanese islands of Honshu and Hokkaido, by crossing under the strait of Tsugaru is a railway tunnel of even greater length and in more difficult ground: it has been under construction since 1964. Both these major projects are more fully described in Volume 1.

In Japan also, the islands of Honshu and Kyushu were linked by the Kanmon tunnel, built 1936–44 as two single track tunnels for 1.07 m gauge track, 3.6 km in length. This has been paralleled in 1970–74 by a new tunnel for the high speed railway, carrying two tracks. The tunnel is 18.7 km long, although the width of the strait is less than 1 km. The deepest point is 67 m below sea level. The undersea section of the tunnel is circular, 10.06 m diameter, with concrete lining 0.9 m thick. The tunnel descends steeply at a gradient of 1.8% to pass under the sea bed with over 20 m cover, and rises more slowly on the Kyushu side at about 0.7%, where it passes under mountainous country and under the built up land between mountain and sea.

Beneath the strait a major fault, including a 30 m wide belt of clay, was encountered, and was overcome by an interesting technique. The first operation was to drill ahead horizontally some 200 holes lined with 110 mm steel tubes at

200-300 mm centres, and to inject grout round the tubes and cement fill them. The tubes constituted buried hoods for the main tunnel and for a series of headings. Excavation and concreting for the walls and roof proceeded successively upwards in these headings from the bases of the walls, the invert being left to the last.

SHIN – KANMON TUNNEL

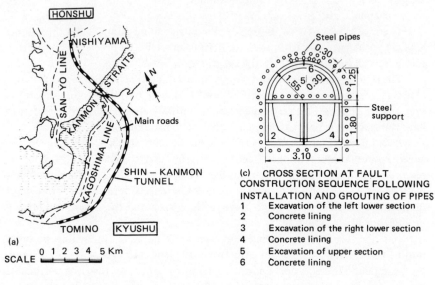

(c)   CROSS SECTION AT FAULT
CONSTRUCTION SEQUENCE FOLLOWING
INSTALLATION AND GROUTING OF PIPES
1   Excavation of the left lower section
2   Concrete lining
3   Excavation of the right lower section
4   Concrete lining
5   Excavation of upper section
6   Concrete lining

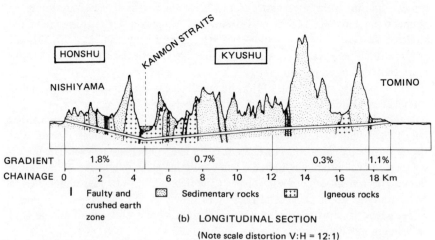

Fig. 8.8 – Shin-Kanmon Tunnel. (a) Plan, (b) Longitudinal section. (c) Special cross section for clay filled fault zone.

Subaqueous tunnels in general are aligned as nearly as practicable to cross the waterway by the shortest route, at the shallowest depth, but geology as well as geometry must be considered, and a longer, deeper route may offer more favourable tunnelling conditions, particularly in respect of favourable rock and inflow of water whether through permeable and jointed rock or at faults.

The questions of alignment, gradients, curves, cross section and lining have already been discussed. In subaqueous tunnels more emphasis is necessarily placed on waterproofing and drainage and ventilation problems.

In such long tunnels, with no intermediate shafts, yet with the necessity for the utmost economy in cross section, aerodynamic problems may be of major importance, particularly if high speed running is required.

8.7 TUNNEL LININGS

There is little about tunnel linings that is special to railways. In the great railway age of the nineteenth century brickwork was the normal lining in British practice. The 'English Method' was based on alternate excavation of a length and building of its brick lining. A merit of brickwork was its adaptability to varying conditions: an extra ring of brickwork could readily be built in the arch if found necessary.

Fig. 8.9 – Old brickwork undergoing repair using quick setting waterproof mortar in London tunnel. (*Ronacrete Ltd.*)

Adequate supply of bricks was a problem in long tunnels, sometimes solved by setting up a brickyard, utilising excavated clay from the tunnel itself or approach cuttings.

In the long term, deterioration of some brickwork linings has resulted from attack on the lime or cement mortar by percolating water, possibly containing sulphates. Acid attack has also come from the products of combustion from steam or diesel locos. Renewal of one, two or more rings of brickwork is possible.

Most other types of lining have been employed. Masonry, where suitable stone and skilled masons could be obtained, has been widely used in the past. *In situ* concrete has been, and is, used on an extensive scale. Sprayed concrete may be adopted either as an immediate lining to be followed by *in situ* concrete, or as a complete lining. In circular tunnels, cast iron segments or precast concrete segments are particularly applicable to subaqueous tunnelling. Steel arch ribs may provide immediate temporary support, but require to be subsequently embedded in *in situ* concrete or sprayed concrete.

Decorative aspects have little relevance to main line railway tunnels, but a smooth finish may be of considerable importance for the aerodynamic aspects of high speed railways.

The importance of waterproofing, particularly in the crown, has already been referred to.

8.8 CURRENT ACTIVITY

Although less in the public eye than highway tunnels there is very considerable world activity in construction and planning of railway tunnels. The Channel Tunnel and Seikan Tunnel have been dealt with in some detail in Volume 1. The Japanese rail system, as already noted, has been and is building numerous mountain tunnels and is stated to have over 3500 tunnels with a total length exceeding 1500 km.

Italy has built the 250 km 'Direttissima' high speed railway between Rome and Florence, incorporating four major tunnels of lengths 5.7, 7.4, 9.3 and 10.9 km and numerous shorter tunnels totalling a further 27 km. Techniques employed have included use of a tunnelling machine and extensive treatment with complex grouts. By way of contrast tunnels have been avoided in the new French line between Paris and Lyon, very deep cuttings being preferred to avoid the aerodynamic pulses occurring with high speed trains in tunnels: gradients up to 3.5% have been employed as being practicable in the absence of heavy goods traffic.

In New Zealand the difficult 8.8 km Kaimai Tunnel was opened in 1978, giving direct rail access to the port of Tauranga. Canada plans long new rail tunnels through the Rockies to cope with the growing volume of bulk haulage. Greece hopes to construct new railway routes through its mountainous terrain, and in Iran plans have been prepared for major works.

Fig. 8.10 – Kaimai Tunnel, New Zealand. (a) Opening September 1978. (*New Zealand Railways.*) (b) Rails in 160 m lengths being transported into tunnel. (*New Zealand Railways*)

In London proposals are under consideration for a deep tunnelled cross city link between south and north. New York's East 63rd Street crossing under the East River includes provision for main line tracks.

In Switzerland the 5 km Heitersberg tunnel has been completed on the Zurich–Berne line and plans have been made for a new St. Gotthard base tunnel which would have a length of 49 km.

Jugoslavia completed in 1976 a railway from Belgrade to the Adriatic coast, through very difficult mountain country, where a 250 km length required 254 tunnels up to 6 km in length and, to reach the 1032 m summit, a 3.2 km tunnel on a gradient of 2.5%.

BIBLIOGRAPHY

This bibliography, though brief, ranges widely in time and space, including mostly descriptions of specific tunnels of special interest.

Kossuth, F., The Mont Cenis tunnel, *Engng*, 1871, **11** pp. 347, 377, 421, 429; **12** pp. 37, 180, 193, 211, 241, 283, 367; 1872, **13** p. 391.

Fox, F., The Mersey Railway, *Min. Proc. Instn Civ. Engrs*, 1886, **86**.

Walker, T. A., *The Severn Tunnel*, Bentley, 1888 (reprinted Kingsmead, 1969).

Rogers, A. E., The location and construction of railway tunnels, *Min. Proc. Instn Civ. Engrs*, 1901, **146**.

Fox, F., The Simplon tunnel, *ibid*, 1907, **168**.

Jacobs, C. M., The Hudson River tunnels of the Hudson and Manhattan railroad, *ibid*, 1910, **181**.

Wilgus, W. J., The Detroit River tunnel, between Detroit, Michigan, and Windsor, Canada, *ibid*, 1911, **185**.

Henderson, B. H., The Transandine railway, *ibid*, 1913, **195**.

Dennis, A. C., Construction methods for Rogers Pass tunnel, *Trans. Amer. Soc. Civ. Engrs*, 1917, **81**.

Bayley, V., The Khyber railway;

Hearn, G., The survey and construction of the Khyber railway, *Min. Proc. Instn Civ. Engrs*, 1926, **222**.

Keays, R. H., Construction methods on the Moffat tunnel, *Trans. Amer. Soc. Civ. Engrs*, 1928, **92**.

Betts, C. A., Completion of Moffat tunnel of Colorado, *ibid*, 1931, **95**.

Kerr, D. J. *et al.*, The eight-mile Cascade tunnel, Great Northern railway: a symposium, *ibid*, 1932, **96**.

Robarts, H. E., Tunnel maintenance, *Instn Civ. Engrs Rly Div.*, 1947 (23).

Champion, C. L., The construction of the trans-Iranian railway, with particular reference to works of Lot 6 (South), *J. Instn Civ. Engrs*, 1947, **29** (Dec.).

Campion, F. E., Part reconstruction of Bo-Peep tunnel at St. Leonards-on-Sea, *ibid*, 1951, **36** (Mar.).

Scott, P. A. and Campbell, J. I., Woodhead new tunnel: construction of a three-mile main double-line railway tunnel, *Proc. Instn Civ. Engrs*, 1954, **3** (Sept.).

Terris, A. K. and Morgan, H. D., New tunnels near Potters Bar in the Eastern region of British Railways, *ibid*, 1961, **18** (Apr.); *see also* discussion, 1962, **21** (Apr.).

Saito, T., Construction of Shin-Kanmon Tunnel, *Civ. Engng in Japan*, 1974, **13**.

Sacrison, H., Flathead Railroad Tunnel, *Proc. Amer. Soc. Civ. Engrs, J. Constr. Divn*, 1971, **97** (Mar.).

Hermans, R. E. *et al.*, The Kaimai Tunnel (9 papers), *Trans. New Zealand Instn Engrs*, 1978, **5** (1).

9

Ventilation and Aerodynamics in Vehicle Tunnels

9.1 HISTORY

The need to provide ventilation for vehicle tunnels arises mainly from the atmospheric pollution caused by the propulsive units and the heat generated. It is only quite exceptionally, as in the proposed Channel tunnel, that the oxygen requirement of the passengers transported becomes significant.

In the 19th century, railway tunnels faced the problem of pollution by smoke from coal burning locomotives, not important in short tunnels on an open main line, but very serious in urban tunnels and long subaqueous or mountain tunnels. Electrification solved some of these problems and made possible the development of metro systems. Oil burning locos also reduced the smoke pollution but produced their own smoke and irritant fumes.

Ventilation of highway tunnels is a 20th century development, the requirement arising initially from the exhaust gases of petrol driven internal combustion engines, and later also from smoke and fumes produced from heavy diesel engined vehicles.

Pollution levels may be controlled by limiting vehicle access to the tunnel and by improvements in engine design and fuels, but tunnel design must still ensure the supply of adequate fresh air to dilute the pollutants to acceptable levels, and the extraction of the polluted air. In addition to this aspect of toxicity, the progressive accumulation of heat in any tunnel network must be examined, but more particularly in metro tunnels, where direct pollution is minimal. Another need in metros is to avoid excessive draughts in stations.

Entirely different phenomena have more recently become significant in railway tunnels, namely the aerodynamic effects of the passage of high speed trains. Such trains interact with the tunnel structures to generate air pressure pulses causing discomfort to passengers. In addition the train, within the constricted area of the tunnel, is subject to aerodynamic drag, which may reduce speed seriously in a long tunnel, or make necessary much higher loco power and give rise to consequent heat problems.

There are thus three principal aspects to consider: pollution in highway tunnels, heating effects and draught relief in metros, aerodynamic effects in high speed rail tunnels.

The need to safeguard users of long highway tunnels became apparent in design of the Holland Tunnel (1927) in New York under the Hudson river. The essence of the solution was the dilution and extraction of the toxic exhaust gases. For many years the proportion of CO in the atmosphere to which vehicle drivers were exposed was seen as the fundamental criterion of any ventilation system, and it remains in most cases the primary consideration, but the importance of other aspects has become apparent. The proportion of diesel engined traffic has since grown to very significant levels. Their emission of CO and other toxic gases is low compared with petrol engines, but they create smoke to a degree that interferes with visibility, and also gases which, although non-toxic, can be unpleasant unless adequately diluted.

From the first the problems of fire control have also been of major import-ance in design of ventilation systems. Another aspect, which normally requires no separate provision, is the heat energy given out by vehicles (1) in the exhaust gases, (2) from radiators and (3) from brakes. Fresh air requirements for dilution of noxious fumes are usually more than adequate to carry away the surplus heat without the temperature of the tunnel structures building up to objectionable levels.

The earlier tunnels, were usually designed to accept two-way traffic either normally or occasionally, but probably a majority of modern highway tunnels carry one-way traffic, except temporarily in emergency or for reasons of main-tenance. There is therefore a traffic-induced flow of air, which may in some cases be wholly adequate for ventilation requirements, and, at least, contributes very substantially. It is in some degree self-regulating, in that more vehicles and higher speeds induce greater flow of air. Conditions with congested traffic moving slowly, or total blockages, remain to be provided for by fan installations.

The principal patterns of ventilation are longitudinal, semi-transverse and fully transverse. These are discussed in more detail below. Design has progressed from the early 'static' concept of dilution and removal of CO, by transverse air flow across the traffic space at each point where it is generated along the tunnel, to more complex concepts making use of the longitudinal flow and turbulent mixing of the air and exhaust gases, resulting from aerodynamic drag of the vehicles.

9.2 HIGHWAY TUNNEL REQUIREMENTS

Artificial ventilation for highway tunnels other than very short tunnels is normally essential, even if only for use at times of traffic congestion. While traffic in a tunnel is one-way and free flowing the traffic-induced air flow is found to be adequate up to a length of 1–2 km, and possibly longer, but if traffic may be stationary or crawling forced ventilation is likely to be essential. 'Natural' ventilation is thus only adequate for one-way tunnels not subject to congestion up to about 0.8 km and for two-way tunnels of half that length.

For the occasional use required in relatively short tunnels jet fans meet most requirements. Although not efficient in terms of energy consumption they are economically advantageous, becuase they do not require major structures to house them, nor tunnelled ducts for supply and exhaust air.

For longer tunnels major fan installations are necessary, and may present difficult design problems involving heavy costs.

The system to be adopted can only be selected after study of every aspect: no single factor can be optimised in isolation. Safety, amenity, aerodynamic efficiency, tunnel construction, maintenance problems, capital costs, operating costs, system of operation and traffic control and pattern and composition of traffic must all be examined in the attempt to achieve a balanced effective and economical system, contributing to the main objective of providing an unobstructed highway.

The fundamental requirements are that the ventilation system should:

(1) dilute and remove toxic gases, in particular carbon monoxide,
(2) dilute and remove smoke, generated largely by diesel engines,
(3) remove the heat energy given out by vehicles,
(4) be adaptable to the needs of fire fighting.

The evaluation of air requirements is a subject of much difficulty in detail because individual vehicles vary greatly in their output of pollutants, and the composition and density of traffic flow fluctuates over a wide range from a nearly empty road to a state of acute congestion.

In a long tunnel the air mass moved by fans is very large and there is a considerable time lag in response to any changes in fan speed. The supply of air cannot be varied with the minute to minute fluctuations of traffic flow. Automatic, or manual, regulation cannot be efficient if it simply follows an indicated level of pollution. Some degree of prediction is almost essential, but can only be based on experience of the particular tunnel.

After examination in detail of all the varying factors the ultimate installation must provide safety in the worst conditions, together with economy in normal operation, having due regard also to amenity.

Standards have to be specified for acceptable levels of pollution in terms of toxicity and visibility and for maximum acceptable air velocity. Energy consumption costs, capital costs and maintenance and supervision costs have then to be assessed in comparing alternative systems.

The most important factors in the assessment of needs are:

(1) length between shafts, and number and direction of lanes,
(2) traffic flow patterns and composition, and traffic control,
(3) environmental impact.

The normal pattern and speed of traffic and the appropriate levels of amenity will be influenced substantially by the location and consequent function of the highway tunnel as, for example: subaqueous, mountain, motorway or urban.

9.3 POLLUTANTS

The common pollutants arising in tunnel conditions from engine exhausts are: carbon monoxide, nitrogen oxides, sulphur dioxide, organic compounds (principally unburnt hydrocarbons and aldehydes), carbon particles (soot), lead compounds.

Published figures for exhaust gas composition vary over a very wide range. The figures below are typical for cruising conditions but higher values are experienced, in acceleration for some pollutants, and in deceleration for others.

Toxic and irritating exhaust gases in parts per million (ppm)				
	CO	$NO + NO_2$	Hydrocarbons	Aldehydes
Petrol Engines	27,000	650	1,000	10
Diesel Engines	trace	240	100	10

The physiological effects of these substances are briefly described below.

9.3.1 Carbon monoxide

The major pollutant from petrol engine exhausts is carbon monoxide whose deadly character at quite low concentrations is well known. It usually constitutes about 3% of the exhaust gas volume and is very rapidly dangerous at a level of 0.1% (1,000 ppm). Dilution of the exhaust fumes by a factor of about 200 is essential, that is to say from 3% to 150 ppm.

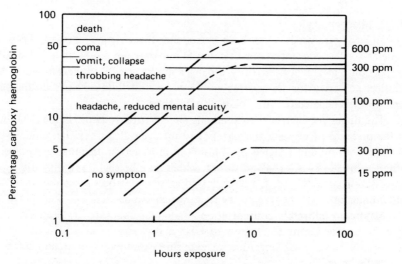

Fig. 9.1 – Carbon monoxide – physiological response of people with normal health, at light activity, to various exposures. (*Pursall and King, Aerodynamics and Ventilation of Vehicle Tunnels, publisher B.H.R.A.*)

CO acts by combining with haemoglobin in the bloodstream, thus depriving the system of oxygen. The effect of exposure can be measured in terms of the percentage of carboxy-haemoglobin in the blood. Below 10% there are normally no obvious effects, but above this level headache, fatigue and dizziness may appear and with rising concentration symptoms will become progressively more severe, with possible collapse at 30%–40%, followed by coma at higher levels and possibly death above 60%–70%

The time taken to reach any level depends on:

(1) initial conditions for the individual (3% would not be unusual)
(2) CO concentration
(3) time of exposure
(4) activity, for which a factor varying from 3 for rest to 11 for heavy work is applicable,
(5) height above sea level.

At sea level a passenger through a tunnel with a concentration as high as 250 ppm requires about two hours exposure to reach 10%, but for someone doing heavy work the time is only about 20 minutes.

At a height of 1500 m these times are reduced to about 1 hour and 10 minutes respectively while at a height of 3000 m the altitude factor alone results in a deficiency exceeding 10%. Special studies have therefore been necessary for such tunnels as the Eisenhower Memorial tunnel through the continental divide in Colorado at a height of 3350 m (11200 ft). At this height air pressure, and therefore density, is only about two thirds of that at sea level. An hourly average for CO not exceeding 100 ppm under adverse traffic conditions was specified for the ventilation system.

9.3.2 Nitrogen oxides

NO and NO_2 are produced in both petrol and diesel engine exhausts. They are toxic at smaller concentrations than is CO but they are produced in yet smaller proportion and are not currently a danger in operational tunnels in which the dilution necessary for CO is effected. They do constitute a hazard during construction when generated from blasting explosives.

The direct product in the exhaust is almost entirely Nitric oxide (NO) but, in the presence of oxygen, this oxidises to nitrogen dioxide (NO_2), which is the more toxic form and which is most insidious in its action. Its inhalation, causing irritation of the eyes, nose and throat, which may be slight and may disappear on cessation of exposure, can be followed after some hours by serious irritation and inflammation of the respiratory tract, which can be fatal.

Maximum allowable concentration for 8-hour exposure has been set at 25 ppm, but more recent regulations in USA and Germany set a limit of 5 ppm. 100 ppm is considered dangerous for exposure exceeding ½ hour, and 200 ppm is 'rapidly lethal'.

One important and dangerous difference from CO is that NO_2 symptoms cease to be reversible beyond some threshold, whereas CO symptoms at almost any stage are reversible by supply of clean air.

Nitric oxide (NO) appears to act in the same way as CO by combination with the haemoglobin in the blood, and although less toxic than NO_2, it is much more toxic that is CO.

9.3.3 Sulphur dioxide
SO_2 is the principal oxide of sulphur experienced. It is a bronchial and nasal irritant with a characteristic taste. It is unpleasant in the range 5 to 10 ppm. The production from motor vehicles is a very small proportion of the total from other sources and it does not normally present any tunnel ventilation problem.

9.3.4 Organic compounds
The production of these is extremely variable. In combination with oxygen and oxides of nitrogen unburned hydrocarbons can give rise to 'photochemical smog' in the presence of ultraviolet light. They are the cause of eye irritation.

Unburned hydrocarbons are mainly attributable to badly tuned engines.

Aldehydes, alcohols and organic acids have been identified in highway tunnels. Although unpleasant odours are observed no health risks have been found. Careful study of the health records of employees in tunnels of the New York Port Authority have shown no special identifiable risk over long periods.

9.3.5 Particulate matter
Carbon, mainly from diesel engines, but also from over-rich mixtures in petrol engines, is the principal solid particle which may act as a nucleus for droplet formation and consequent haze. Very many other solids have been recorded.

9.3.6 Lead
The use of lead to improve octane ratings constitutes a continuous source of compounds with oxygen, chlorine and sulphur. About 40% of the lead in the fuel becomes airborne. Lead is a cumulative poison, slow in action. It is not considered to be at a dangerous level in tunnels but reduction of permissible levels for lead additives in petrol is being effected in most countries, more particularly in Germany. The possibility that air exhausted from road tunnels may cause undesirable concentration of lead in areas near the discharge shaft has recently emerged as a potential hazard to be studied and avoided. At present it seems likely that the problem is limited to areas where lead is already present in quantities which make further accumulation unacceptable.

9.3.7 Acceptable standards for toxic substances
There are no statutory standards for pollution directly applicable to tunnel atmospheres. The responsible authority promoting any tunnel project will be guided by reports and studies on a world-wide basis.

Threshold Limit Values (TLV) recommended for various airborne noxious substances are published by H. M. Factory Inspectorate, based on a list adopted by the American Conference of Government Industrial Hygienists in 1968. They are more particularly related to exposure in factory conditions throughout an 8-hour day and a 5-day week and are not directly applicable to tunnel conditions, but do offer helpful guidance. Higher values are in most cases acceptable for short term and intermittent exposure.

The relative significance of the various pollutants actually experienced in road tunnels may be put in perspective by a comparison of these TLV figures with maximum figures measured in a series of Drager Tube samples in a number of UK tunnels.

Substance		TLV	Max. measured
Carbon Monoxide	CO	50 ppm	110 ppm
Nitric Oxide	NO	25 ppm	5 ppm
Nitrogen Dioxide	NO_2	5 ppm	0.25 ppm
Sulphur Dioxide	SO_2	5 ppm	0.5 ppm
Hydrocarbons		111 mg/m^3	10 mg/m^3
Carbon Dioxide	CO_2	0.5%	0.05%
Lead	Pb	0.15 mg/m^3	0.032 mg/m^3

It is clear that in practice CO is fundamental in determining the necessary degree of dilution as none of the other pollutants are observed at more than 20% of TLV. Oxides of nitrogen, however, call for a special note of caution for the reasons already discussed and in the light of suggested lower limits proposed in recent years for these substances. Questions of smoke and visibility are separately discussed.

9.3.8 Limits for carbon monoxide

Acceptable standards for CO require to be examined in more detail. The fact that the TLV of 50 ppm is substantially exceeded in normal practice is entirely proper because the exposure of drivers and passengers is for a very short time only.

In the first of the modern highway tunnels, the Holland Tunnel in New York (1927), the design maximum for CO was 400 ppm, with the assumption that the average would be much lower. Even this value is unlikely to have serious consequences for brief exposures, but an emergency maximum of 250 ppm or 200 ppm, with a working level not exceeding 100 ppm or 150 ppm is more usual now. It is of interest to note that the mean level in the Holland Tunnel has in fact remained at about 70 ppm over the half century of its use.

The Permanent International Association of Road Congresses (PIARC) through its Technical Committee Report on Road Tunnels presented at the 16th Congress in Vienna in 1979 has recommended that ventilation should be so designed that the average CO level at any time in normal conditions should not exceed 150 ppm, with an absolute limit of 250 ppm for accidental and short peaks anywhere. With longitudinal ventilation, in which there is a maximum of twice the average at the point of discharge, 250 ppm maximum is considered acceptable.

9.4 AIR SUPPLY

Whatever limit is adopted there will inevitably be great short term variation in the levels of CO actually experienced, because air supply cannot be regulated to follow immediately and closely fluctuations of traffic in density and composition.

9.4.1 Dilution of CO

The basic formula for air requirements for a single lane in terms of dilution of CO is:

$$Q = \frac{fq}{60S} \times \frac{10^6}{\delta} \ \text{m}^3/\text{s.km}$$

where f = coefficient for gradient and altitude
$\quad\quad q$ = unit volume (l/m) of CO emitted by a typical vehicle
$\quad\quad S$ = average spacing of vehicles (m)
$\quad\quad \delta$ = permitted CO content of air in parts per million

Diesel engined vehicles are intially equated to cars although their percentage of CO in the exhaust is much lower, offset in part by the higher volume of exhaust from larger engines. Smoke output needs to be examined separately if the proportion of diesel vehicles is other than small.

The value for δ in this formula has already been discussed: f, q and S require examination.

9.4.1.1 Gradients and altitude

The appropriate corrective coefficients for gradient are in some degree dependent on the composition of traffic. For a mix of 90% cars and 10% heavy diesels the recommended coefficient varies from 0.7 for a 6% descending gradient to 1.3 for a 6% ascent. If the tunnel is two-way these tend to counterbalance and the modification is not of great significance, but in a one-way tunnel they must be taken into account particularly on a continuous rising gradient.

They may also be important in a subaqueous tunnel where the profile includes a descent, a mid-river level length and an ascent, the gradients being typically of the order of 4% to minimise the tunnelled length. The rising gradient may slow down heavy vehicles disproportionately and some congestion may result, contributing further to pollution. Local conditions over a short length of tunnel are more immediately relevant where the ventilation system is fully transverse, because any longitudinal element in the air flow tends to average out pollution. On the other hand, a heavy vehicle travelling at the same speed as the longitudinal air flow will tend to carry with it its own exhaust output.

At high altitudes two direct effects must be taken into account: increased CO generation because of reduced engine efficiency, and increased volume of air necessary because of reduction in density. These are combined to give a height correction coefficient ranging from 1.0 at sea level to 2.9 at 2000 m. There is also a physiological effect because of the oxygen deficiency.

9.4.1.2 CO Emission

In determining the air supply necessary to reduce to acceptable dilution carbon monoxide, or any other pollutant, it is essential to ascertain the characteristics of individual vehicles and to establish a typical basic emission. The output of CO from the average vehicle can only be an approximate estimate. It obviously depends on the engine capacity and on its rate of consumption of fuel, but also depends on the design and tuning of the engine and on its speed, and whether accelerating, decelerating or running at a steady speed. Outputs up to 40 l/min have been quoted for level roads, averaged over various speeds and cycles of operation, and up to 80 l/min for ascending gradients. Even higher figures may apply for acceleration, but these are not relevant to sustained speeds in a tunnel. A French survey in 1967 gave observed average rates varying from 15 to 28 l/min for tunnels in Holland, France and Switzerland.

The PIARC Technical Committee in 1979 published a new basic emission curve from a typical vehicle. The output rises from 8 l/min when idling to 16 l/min at 20 km/hr, remaining constant at high speeds.

The former curve (of 1975) continued to rise to a maximum of 22 l/min at 80 km/hr, but a more significant change has been the elimination of a peak of 22 l/min at 10 km/hr for very congested conditions, in which vehicles are moving intermittently and accelerating in low gear. These changes are attributed to improvements in engine design and control of emission, and also to new studies of traffic in stop/start conditions.

In using any such curve for air requirements it must be combined with an estimate of vehicle spacing

9.4.1.3 Traffic density

The concept most familiar to traffic engineers is that of vehicles per hour per lane, but this must be translated into vehicle spacing, or traffic density, for use in calculations of tunnel pollution. The quantities are related by the equations

$$ S = \frac{10^3}{n} = \frac{V \times 10^3}{N} $$

where S = spacing of vehicles centre to centre (m)
  n = vehicles per km = traffic density
  V = average speed (km/hr)
  N = traffic flow in vehicles per hour

What is found in practice is that vehicles space themselves out in accordance with the speed of the traffic flow, more particularly in tunnels in which overtaking is not permitted. When a line of traffic is stationary the minimum spacing is about 5 m for cars. In motion this opens out immediately to about 10 m or more, and increases progressively with speed, provided that a condition of free flow is established. If there is further obstruction, the spacing again decreases towards complete congestion; under these conditions there is no stable relationship between speed, spacing and pollutant output. Speed and spacing vary erratically in stop-start motion, and pollutant output fluctuates.

If traffic is flowing freely a rate of 2000 vehicle/hr or more for a single lane may be achieved, but maximum normal sustained flow is about 1750 vehicle/hr anywhere in the speed range 25 to 55 km/hr. This corresponds to spacing from 14 to 31 m. Above 55 km/hr the capacity tends to become less because of greatly increased spacing.

These generalisations need to be examined more closely for the traffic in any particular tunnel. In mountain or motorway tunnels congestion is unlikely except as an occasional emergency, but in urban tunnels unobstructed conditions of exit may be of fundamental importance if congestion is to be avoided. On the other hand if entry to the tunnel can be restricted by traffic control signals the congestions of peak hour traffic may be kept outside in free air.

In subaqueous tunnels a characteristic feature tending to cause congestion is that a descent and a level length are followed by a rising gradient which tends to reduce abruptly the speed of vehicles, particularly those heavily loaded and requiring the use of a low gear. A form of 'shock wave' in the traffic flow has been observed, but studies in New York tunnels demonstrated the possibility of improving traffic flow in those conditions by controlled admission, limited to the current capacity of the lane.

9.4.1.4 Air requirements

In respect of CO there are thus three speed ranges to examine in providing ventilation capacity: namely, stationary traffic, congested flow up to about 20 km/hr and free flow from 25 km/hr upwards.

First, when traffic is at a standstill, whether from congestion ahead or from an accident or emergency the spacing may be reduced to 5 m. With engines idling the suggested CO output is 8 l/min per vehicle. The air requirement for one lane with a limit of 150 ppm is then 178 m^3/s km but with the limit increased to 250 ppm this becomes 107 m^3/s. To the extent that emergency instructions to switch off engines can be enforced, pollution can be reduced.

The second range is that of congested traffic moving intermittently. If such a condition is allowed to continue the worst case might be supposed to be progress at 10 km/hr with a spacing of 10 m, which implies a traffic flow of 1000 vehicles/hr. The suggested CO output is about 14 l/min which gives an air requirement of 156 m³/s.km if 150 ppm is to be achieved or 94 m³/s.km if 250 ppm is acceptable as an occasional short term condition. These figures are rather uncertain in that the average spacing over 1 km is unlikely to be as close as 10 m in the stop-start motion but the output of 14 l/min may be exceeded locally.

The third range for steady free flow at 25 km/hr upwards requires the maximum capacity at any speed to be combined with the output of 16 l/min per vehicle. Taking this capacity as 1750 vehicle/hr the spacings at 25 and 55 km/hr are respectively 14 and 31 m and the air requirements are 127 and 57 m³/s km for a pollution level of 150 ppm.

It is obvious that the air requirements decrease progressively and substantially as average speed increases. It must be noted that they are calculated for ordinary maximum capacity at the speed selected but that traffic density may often be much lower for long periods, and air requirements will be lower in proportion.

A supply of about 150 m³/s km is thus likely to be adequate as a basis for a normal tunnel, provided that it is available quickly if congestion develops.

9.4.2 Diesel engined vehicles

The calculation of requirements arising from diesel engined vehicles is less direct. Carbon monoxide and other gaseous pollution from this source is adequately diluted without separate assessment, but smoke, principally comprised of solid particles of finely divided carbon, obscures visibility in the tunnel atmosphere, and its dilution to acceptable levels must be ensured. Additional provision is not likely to be necessary unless the proportion of heavy diesel vehicles is over about 15%, or on rising gradients where maximum power can only be achieved at the expense of smoke emission particularly in the case of underpowered and over-loaded vehicles in poor mechanical condition.

9.4.2.1 Visibility limits

The fundamental principle is that of ensuring visibility adequate for safe driving by specifying an appropriate coefficient of absorption defining the rate of loss of light. The standard of lighting must be considered and the safe stopping distance for vehicles, which depends on speed and gradient and also on vehicle design and maintenance.

Direct measurement of loss of light along a set length of the tunnel can be used to assess visibility. In some countries control is based on the mass of airborne soot particles, but this is less direct and is complicated by the presence of particles of road dust, which may contribute a substantial proportion of the suspended mass but, being less finely divided, cause little obscuration.

The PIARC Technical Committee (1979) has published a formula closely analogous to that for carbon monoxide assessment and a series of curves relating to diesel smoke. The basic curve relates obscuration effect of exhaust smoke to vehicle speed the value shown being then multiplied by vehicle weight and by appropriate factors for gradient and altitude, and by the number of such vehicles per kilometre in order to arrive at the air supply required.

It is interesting to note the suggested dilution of exhaust fumes is of the order of 200:1, very similar to the figure for 3% CO in car exhausts, but the volume per vehicle is higher, while the number of vehicles per km is lower.

9.4.3 Total air requirements

The design must provide enough air for safety, not necessarily for comfort, in the worst conditions forseeable. This may be the volume of air necessary to keep CO pollution in an emergency traffic jam below some such level as 250 ppm. The fan capacity and power for this should be quickly and reliably available with standby provision to cover breakdowns and maintenance operations. It may be noted that CO pollution and obscuration by smoke are independent phenomena and their air requirements are not additive: the same air can do both jobs.

The ordinary hour-to-hour air requirements will be substantially less than the emergency peak, and economy in operation will require a system of control and regulation of fan use, possibly combined with traffic control.

Short tunnels in which one-way traffic flows freely are largely self-ventilating, but in urban conditions may be subject to peak hour congestion, requiring ventilating fans.

A long tunnel exceeding 1 or 2 km is unlikely to be self ventilating unless conditions are unusually favourable and must therefore be provided with fans for continuous operation. Substantial economies can be effected by regulating the fan output from zero or a minimum at night to a normal maximum at peak traffic periods and possibly a further boost in emergencies. This is more fully discussed below.

It is not unlikely that CO emissions will in the future be appreciably reduced by the influence of regulations for vehicle manufacture, but requirements for fresh air may not be correspondingly diminished, because higher standards for air purity may be applied and also because of the visibility standards. Even if perfect combustion, or electric propulsion, became practicable any great reductions in air supply would give rise to problems of the build up of heat in the tunnel.

Intake air temperature can be significant, especially a high average over a long perid. There is considerable transfer of heat between the air and the tunnel walls resulting in fairly steady temperatures at mid-tunnel. Close to the intakes, however, it may be necessary to safeguard water pipes against freezing, and in warm humid conditions condensation of moisture from the air may be objectionable. In very hot climatic conditions the tunnel temperature could become unacceptably high as the energy from fuel consumed in vehicles is all ultimately transferred as heat into the tunnel air. The heating problem is more fully discussed below in the context of metro tunnels. It has been estimated that heat balance alone might require air supply at about half the current rates determined by pollution.

Another point to be noted in determining the volume of air required is its purity. It should be drawn from a source as free as practicable from pollution. In city streets CO can exceptionally be as high as 50 ppm: twice as great a volume of such air would be required to achieve dilution to 100 ppm within the tunnel. Factory chimneys in the vicinity may also contribute dust, smoke, fumes or other pollutants particularly when the wind is in certain directions.

In practice the normal provision for major urban tunnels, and others liable to congestion, lies in the range 100 to 200 m³/s km per lane and usually in the upper half of that range. In some, particularly where a fully transverse system is used, the supply may be varied in accordance with the gradient. Examples are: Mersey Kingsway 155 Elbtunnel (Hamburg) 165-185, Coentunnel (Antwerp) 175, Lincoln Tunnel (NY) 114-248, Vieux Port (Marseilles) 125-250. It is found in practice that the full supply is not required except for short peak periods, and that substantially lower levels of supply are adequate for much of the time.

In contrast, the long Alpine and other mountain tunnels are generally designed for traffic much below road saturation capacity. The Mont Blanc Tunnel, 11600 m long at a height of approximately 1300 m, was originally designed for a supply of only 26 m³/s km per lane on an assumed limited traffic flow of 400 vehicles/hr. In 1970 the average flow was 2067 veh/day but with a maximum of 7652 per day and 762 in one hour. Because of increasing traffic the fresh air supply is accordingly being increased by 50%. Other Swiss mountain tunnels have been designed within the range of 47 m³/s km per lane for a 2.5% descending gradient to 154 m³/s km per lane for an ascending gradient of 2.6%. The Eisenhower Tunnel at a height of 3350 m (see above, p. 240) has a provision of 280 m³/s km per lane to meet the needs of heavy traffic flows in the summer holiday period. In Austria the more recent motorway tunnels such as the Arlberg allow 135-190 m³/s km per lane.

9.5 SYSTEMS OF AIR DISTRIBUTION

The conventional classification of systems as longitudinal, transverse, and semi-transverse is an over simplification.

The simplest form is purely longitudinal ventilation, in which the air enters at one portal and is discharged at the other. The movement may be maintained by wind, and more rarely by pressure differences, or by the drag of vehicles in a one-way tunnel, or by a fan or jet system.

Purely transverse ventilation is independent of any longitudinal flow and depends on flow across the tunnel from openings in a supply duct to openings in an extract duct, both ducts following the whole length of the tunnel and fan operated, but in one way tunnels traffic flow contributes a longitudinal element.

Semi-transverse ventilation typically comprises a fan operated supply duct as for the transverse system, but no separate exhaust duct, the whole traffic space functioning for that purpose and discharging either at a portal or through an extract shaft.

In any system of forced ventilation the energy considerations are significant, and for long tunnels can be of major importance. To supply a tunnel of length, L, through a duct of length, l, and cross-sectional area, A, one of the biggest elements of the power requirement is the duct loss, which is approximately proportional to:

$$fLl^3/A^{2.5} \quad f \text{ being a friction factor.}$$

It is obvious from this formula that for maximum economy duct lengths must be short and areas large. Tunnel length, L, cannot be altered except by dividing into sections; duct length, l, must be kept to a minimum; duct area, A is usually severely limited by tunnelling considerations.

Energy advantages for the longitudinal and semi-transverse systems arise from the use of the whole traffic space as a principal air duct.

9.5.1 Longitudinal ventilation

Very short tunnels, particularly in mountain areas where wind is usual and in which traffic is uncongested, are normally self-ventilating by longitudinal flow. An extreme example, which cannot be called short, is the Tateyama Tunnel in Japan, which is a single lane tunnel of which the main section is 2100 m in length, passing under a ridge at a height of 2400 m. Natural ventilation was found to be adequate, but the designed capacity is only 24 cars per hour.

With one-way traffic in a tunnel the vehicles can induce air flow with a velocity of the order of one third of the traffic speed. The adequacy of the air supply then depends on maintaining traffic speed and avoiding congestion, which is doubly detrimental in that air flow is reduced at the same time as traffic density is increased. Unless traffic can be very precisely regulated as to numbers and speed, any such tunnel will require at least standby mechanical ventilation for use in times of congestion.

The principal advantages of longitudinal ventilation are simplicity and minimum energy demand, both reflected in low capital and operating cost. The disadvantages appear progressively with length of tunnel and volume of traffic, as more air can only be provided by increased velocity of flow.

Up to 10 m/s in the direction of traffic flow can be accepted with one-way working, but with two-way traffic a normal limit might be about 5 m/s (18 km/hr) or exceptionally 7 m/s. If a maximum air supply is specified at the rate of 150 m³/s.km and the traffic lane dimensions including clearances are 4 m wide × 5.5 m high the velocity and length are proportional. An air speed of

7 m/s limits the length to 1000 m. Reduced supply at 100 m^3/s.km would allow the length to be 1500 m.

Higher velocities have been adopted for some of the long alpine tunnels.

The risk that fire will spread rapidly along a congested tunnel, and difficulties in the control of smoke must be fully considered in application of the purely longitudinal system.

With longitudinal flow the degree of pollution rises uniformly along the length of the tunnel to the maximum acceptable at the point of discharge, which is normally the portal, but may be an extract shaft. As average pollution is only half the maximum there is a case for allowing a relatively high peak value, unless stationary vehicles are to be expected at the maximum point. The P.I.A.R.C. standards take account of this.

The polluted air is most readily discharged directly through the tunnel portal, but this may be unacceptable in urban conditions in the immediate vicinity of public or commercial buildings or in residential areas. In such cases an extract fan and discharge shaft may become obligatory. Another aspect may be the risk of recirculation of the polluted discharge into an adjacent tunnel entrance.

It is only for short lengths and free flowing one-way traffic that traffic-induced air flow can be adequate. Forced ventilation becomes essential either because of length of tunnel or congested traffic. The fan installation may be a boost system within the tunnel or may comprise a shaft for fresh air supply, a duct and guide vanes into the tunnel, and a corresponding extract system.

Either system may be used to supplement or replace traffic-induced air flow; with the latter it is possible to ventilate a long tunnel by providing intermediate shafts which divide it into sections.

Many short urban tunnels, as for underpasses, are likely to be self-ventilating for long periods when traffic is flowing freely but can lose almost all their self ventilating capacity at rush hours when traffic is blocked. Boost fans mounted outside the traffic space, near the roof of the tunnel can be used to provide the necessary flow. In energy consumption they are relatively inefficient, but they are in use only for limited periods and are cheap to install and maintain, normally requiring no elaborate ducting.

Tunnels of circular or horseshoe cross section provide ample overhead space for installation of small diameter fans for this purpose, but rectangular tunnels, which are the normal type in shallow cut-and-cover construction and are also a common form of submerged tube tunnel, may not lend themselves so readily to the system, although space can be provided, either above walkways, or by local enlargements.

9.5.2 Transverse ventilation

Fully transverse ventilation is probably the ideal system for the central sections of a long tunnel if costs of construction and operation do not preclude it. The necessary air is fed from a duct into the tunnel at or near road level, and the same quantity is extracted at roof level or diagonally opposite. The air movements in the traffic space thus impose no limitations on the length of tunnel and there is no progressive build-up of pollution as in the case of longitudinal flow. Local requirements, as for example on gradients, can be met by increased supply and exhaust at that section, and in the event of fire, smoke is extracted directly without longitudinal spread. The traffic flow may be in either direction, or two-way, without disrupting the pattern of ventilation.

The principal obstacle to the adoption of the system is cost, both in construction and in operation. Ducts must be provided for both supply and exhaust and, as is clear from the equation on p. 249 above, energy costs increase in proportion to the cube of the duct length, and are duplicated for supply and exhaust, while the capital cost of providing in tunnel ducts of large area can be prohibitive. There is also the problem of cleaning and maintenance of the exhaust duct in which deposits of oily soot tend to accumulate.

In a tunnel of circular section there is normally available useful but limited space both below the road level and above the height requirement, and this shape lends itself to fully transverse ventilation, which was adopted for the first of the urban tunnels designed for modern motor traffic, the New York Holland Tunnel (1927). The system has worked well and proved adaptable to developments over nearly 50 years. It performed satisfactorily in a serious fire in 1949. Similar systems were provided by the same authority for the three tubes of the Lincoln Tunnel (1937, 1945, 1957).

Fully transverse ventilation systems were also adopted in the long Swiss tunnels of St. Gotthard (1980), which is divided into five sections, and San Bernardino, and in Holland for the Maas, Velsen, and Ij Amsterdam submerged tube tunnels.

Transverse ventilation systems may be further classified as upward, lateral, etc. according to the positions of the supply and exhaust apertures. The upward transverse is possibly the best because the polluted air is hot and tends to rise and assist the circulation, but the turbulence induced by traffic flow is likely to be dominant. The lateral transverse is suitable for submerged tube tunnels where depth of construction must be kept to a minimum but extra width is more readily available. In some of the horseshoe mountain tunnels both supply and exhaust ducts are in the crown of the tunnel, making it necessary to feed fresh air by local ducts down the side of the tunnel to kerb level.

9.5.3 Semi-transverse ventilation

The semi-transverse systems are, like most engineering designs, a compromise between the ideal, the practical, and the economical. The Mersey Queensway Tunnel (1933) was an early example of such a system in modern conditions. The system was adopted after study of the Holland Tunnel system and after extensive experiments including smoke extraction from fires. Here also the system has stood the test of time and has been developed and adopted in the new Mersey Kingsway Tunnel (1971) and also in the Dartford, Tyne and new Blackwall tunnels. It has also been chosen for similar subaqueous crossings in the Hong Kong Harbour Tunnel and the Vieux Port Marseilles, and in mountain tunnels such as the Scheiteltunnel in Austria and the Mosi Baregg tunnels in Switzerland.

The typical feature is the use of the whole traffic space as the exhaust duct. In the subaqueous tunnels of circular section a ventilation station on each side of the channel supplies fresh air through the space under the roadway and this is fed into the traffic space by regulating apertures at kerb level. At mid-river there is minimal longitudinal flow and from that point back to the exhaust shafts on each side the air velocity in the traffic space increases progressively to a maximum. Maximum acceptable air velocities both in ducts and in traffic space impose limitations on length of tunnel ventilation from a single station, just as in the case of longitudinal ventilation.

The apertures must be carefully calculated to provide uniform supply as the pressure and velocity in the supply duct diminish along its length. The theoretical effect of this system is that with uniform output of pollutants and perfect mixing along the length of the tunnel there will be uniform dilution and therefore a constant level of pollution. Figure 9.2 illustrates the patterns of air flow and pollution for various systems.

By operation of exhaust fans only, it is possible to make a semi-transverse system function as a longitudinal system. It is worth noting that the semi-transverse system comprises transverse supply and longitudinal exhaust, and does not lend itself to the reverse pattern of longitudinal supply with transverse exhaust into a duct, because air exhausted near the point of supply will be almost unpolluted while the total air flow in the traffic space will be steadily diminishing in volume along its length, becoming progressively more polluted to a final point of nil flow and undiluted exhaust fumes.

9.5.4 Mixed systems

Although most tunnels fall broadly into one of the above classes there is often a considerable departure from the simple system such as the combination of semi-transverse and transverse or longitudinal and semi-transverse systems.

The former is illustrated by the initial scheme for the Mont Blanc Tunnel (1965) which is 11.6 km in length and where intermediate shafts were impracticable; the system adopted was effectively transverse at the centre of the tunnel and semi-transverse at the ends. All the ducts are below the roadway so that the crown of the arch is unobstructed and is readily accessible for inspection and treatment of any seepage of water or other maintenance problems. The ventilation design was based on a traffic flow of 415 vehicles per hour and a maximum CO content of 100 ppm.

With great increases in traffic, both tourist traffic up to 800 vehicle/hr in

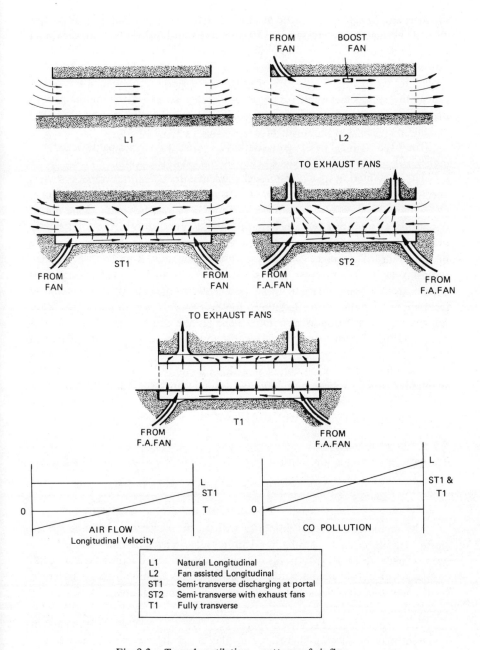

Fig. 9.2 – Tunnel ventilation – patterns of air flow.

summer, and heavy diesel traffic at other times exceeding 60 veh/hr, the acceptable CO average was increased to 150 ppm but additional ventilation was found necessary not only for CO but for visibility when diesel traffic exceeded 60 veh/hr. The scheme adopted was the reversal of the separate exhaust duct to allow supply to be increased by 50%, making the system entirely semi-transverse, except that a higher rate of supply is given progressively in the middle part of the tunnel. In the event of fire the original exhaust system can be used. The maximum air speed in the traffic area is increased from 6 to 9 m/s.

The Clyde Tunnel in Glasgow also has an exhaust system capable of extracting 50% of the supply, the balance being discharged at the portals.

The introduction of a longitudinal element into a semi-transverse system is exemplified in the mid-river section (1214 m) of the Mersey Kingsway Tunnel (1971) where the otherwise uniform supply of air along the length of the tunnel is boosted at mid-river by concentrating about one eighth of the total supply over a 60 m length. This introduces a longitudinal element into the flow, primarily to accelerate the clearance of fumes from the slow moving section where longitudinal velocity is otherwise very small. The calculated time of passage of a particle from a point adjacent to the dead centre to the exhaust shaft is reduced from about 21 minutes to 11 minutes by this concentration of part of the flow. Longitudinal ventilation is also introduced in the short sections of Dartford Tunnel (1963) between portal and ventilation shaft, fresh air being drawn in without any separate fresh air supply duct.

The Elb-tunnel (1975) comprising three 2-lane tubes is equipped with an adaptable ventilation system, functioning as semi-transverse in normal conditions but with extract capacity of 65% of the supply to make it function transversely in congested conditions, and with special provision for fires.

9.5.5 Energy requirements
The basic energy equation for the supply of ventilating air by means of a fan is

$$\text{Energy} = \frac{\text{Volume of air per second} \times \text{total pressure}}{\text{efficiency}}$$

The required volume of air varies with traffic density, rising in congested traffic conditions to something like 150 m^3/s.km per traffic lane. The pressure head is made up of velocity head, frictional resistance in ducts and shock losses at changes of section and apertures. The major energy requirements are governed by the necessary lengths of ducts and their cross sectional areas, for which the equation in para. 9.7 applies.

The magnitudes and relative importance of resistances in the various parts of a system may be illustrated by quoting the design figures for one half of one tube of the Mersey Kingsway Tunnel, where a single axial flow fan having a diameter of 5.2 m supplies 350 m^3/s to 1122 m of two-lane tunnel against a pressure of 695 N/m^2 (2.79 in w.g.)

a

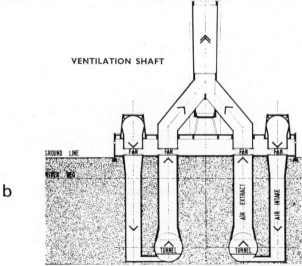

VENTILATION SHAFT

GROUND LINE

RIVER BED

AIR-EXTRACT

AIR-INTAKE

FAN   FAN   FAN   FAN

TUNNEL   TUNNEL

b

Fig. 9.3 – Mersey Kingsway Tunnel ventilation building: (a) As seen from R. Mersey. Note separate louvred intakes for twin tunnels and single exhaust shaft 53 m high. (b) Diagram of ventilation station and shafts.

Source of resistance	Pressure (N/m^2)	%
Intake louvres and duct, fan, safety screen, and shaft and duct into tunnel	204	28
Provision for acoustic treatment	(40)	(5)
Duct under roadway (Area 12 m^2)	419	57
Adjustable slots from duct into road space	37	5
Miscellaneous losses	35	5
	735	100

Insertion of the figure 695 into the energy equation, with an efficiency of 75% gives:

$$\frac{350 \text{ m}^3/\text{s} \times 695 \text{ N/m}^2}{0.75} = 324 \text{ kW.}$$

These resistances are for maximum air flow, but for other flows are proportional to the square of the velocity, energy being proportional to the cube of the velocity.

The largest element is the duct resistance, and the importance of length, area, and smoothness are apparent. The greatest possible duct area, uniformity of section, large radius bends aided if necessary by guide vanes must all be provided in design of the system, together with careful siting of any essential cables and pipes so as to minimise obstruction.

The economy of power in the use of the semi-transverse system is illustrated by comparing the corresponding figure for total pressure in the exhaust system, which is about 27% of the supply pressure. The total power requirement for the semi-transverse system is thus about 1.3 times the power for the supply fan alone, whereas for a fully transverse system, the exhaust duct area would be less than the 12 m^2 of the supply duct and the total power would substantially exceed twice the supply power.

Energy consumption is proportional to the cube of the rate of air supply, and it follows that in large system it becomes essential to provide for reduction when traffic is light. The simplest acceptable pattern would operate with maximum, normal and minimum air flows but more steps are likely to be advantageous in practice. Air supply rates in the ratios 1:2:3:4 will give energy consumptions of approximately 1:8:27:64. The economy of utilising the lowest practicable rate is obvious. On the other hand too complex a system of regulation of flow is not justified because traffic needs cannot be followed quickly and with precision, and too much elaboration may defeat its own ends.

9.5.6 Choice of plant

For major tunnels very large fan capacity is required and fans are normally specially designed for the duty required. A choice must be made between a few large fans and a larger number of smaller units and also between centrifugal and axial flow types. It is not possible here to do more than indicate the broad basis of choice. Local electrical and mechanical facilities are very relevant.

With large fans the desired variability of output may be obtained by multiple pole motors having a range of speeds, or by other electrical devices such as commutator motors or thyristor control, or by gearing or variable fluid couplings.

When a single large fan supplies a section of tunnel the ducting can be designed for the smoothest flow and consequently minimum losses, but it is vital that air supply to the tunnel shall not be interrupted by breakdown of the fan system, and a standby system must be provided to operate very rapidly in emergency. In the Mersey Kingsway Tunnel and in other U.K. subaqueous tunnels 100% standby is made available by provision of a duplicate fan and motor so mounted on a transfer carriage that changeover can be effected in a very brief period. There is also provision against electrical supply failure by means of standby diesel generating plant. The alternative of fixed fans and branching ducts complicates the air flow pattern and results in losses at junctions and discontinuities, as well as introducing further mechanical problems of installation, operation and maintenance.

If a number of smaller fans are installed the complexity of the ducting increases as do the losses, but it may be possible to use standard sizes and types of fans and there is also unlikely to be a need for 100% standby. The air loss resulting from breakdown of one out of three or four fans working in parallel is not catastrophic, and it is likely to be satisfactory if replacement fans and motors, held in reserve, can be installed in the course of a few hours. This basis was adopted for the Hong Kong Tunnel.

A further consideration in making the choice is the accessibility of the site for installation, maintenance and replacement of heavy plant. The differences between a high mountain tunnel and an urban subaqueous tunnel are obvious.

9.5.6.1 Fan type

The principal merits of the axial flow fan are: (1) Efficiency at low pressures, but over a limited range of pressure. (2) Economy of space when installed as a single fan directly in the intake duct.

The radial flow fan operates more efficiently at somewhat higher pressures and is tolerant of a greater range of pressure without substantial loss of efficiency. It is also claimed to be quieter, but when operated at low speed can produce a deep beat sometimes found more unpleasant. More space is required by radial flow fans of equivalent capacity, and the problem of maintaining a clean motor house is more difficult.

Direct cost of fan and motor may be somewhat less for the radial flow fan, but the extra space, and consequently larger building will more than counter-balance this particularly in urban areas.

Radial flow fans are installed in the Holland and Lincoln Tunnels in New York and also in the original Mersey Queensway Tunnel, and the Mont Blanc Tunnel but axial flow fans have been adopted in the Dartford, Tyne, and Mersey Kingsway Tunnels and also in Hong Kong.

Where longitudinal ventilation, with separate shafts for supply and exhaust by fan, is adopted the same considerations apply, but where there are no separate shafts and air enters at one portal and discharges at the other, boost fans within the main tunnel space are necessary. Jet fans are used, mounted high in the tunnel and may be required at intervals along the length.

In a study for a two-lane rectangular tube the National Engineering Laboratory at East Kilbride found that two booster fans of 1 m diameter mounted in the upper corners of the tunnel were required at about 50 m intervals to maintain an air velocity of 6 m/s necessary to ensure that CO was kept below 250 ppm with congested traffic in a tunnel 700 m long.

9.5.7 Regulation and control

From what has already been said it will be appreciated that by proper regulation of ventilation substantial savings of power can be made, but the scope for saving must not be exaggerated. In a typical subaqueous tunnel the annual average power consumption per lane for the passage of 3 million vehicles, may be about 0.1 kWh per vehicle. Obviously doubling an already adequate air supply and increasing the power cost eight times would be absurd, but on the other hand reduction of supply to a point where the public are inconvenienced and, in greater degree, the tunnel staff, is not justified by the much more limited scope for cost reduction.

'Fine tuning' of the air supply to follow a particular pollution standard by monitoring CO in the vitiated air is impossible, particularly in a long tunnel. The CO produced in the tunnel will not reach the gauges and be recorded for some minutes after it is produced. If the fans are immediately speeded up to meet the new situation further time will elapse, so that the increased air supply entering the traffic space is not effective until possibly 15 or 20 minutes after the traffic initiating the change has passed. The inertia of the system may be illustrated by the fact that maximum air supply to a two-lane tube of the Mersey Tunnel is 50 tonnes per minute. It seems clear that regulation of air supply should be in terms of anticipated and actual traffic flow.

It is of interest to note that in the New York Port Authority's Holland and Lincoln Tunnels: 'The main feature of the ventilation control systems now being developed for these tunnels is to base the ventilation decisions primarily on knowledge of actual traffic conditions throughout the various tunnel sections, rather than primarily on measurement of air quality. By estimating emissions from the known traffic, the control system should be able to anticipate the need for more or less ventilation, and to act accordingly so the effect of the decision will be felt closer in time to the need for such response'. (BHRA B3-37).

The daily pattern of traffic flow is known for most tunnels, building up in urban tunnels from zero at night to a morning peak, falling off again until an evening peak period and thereafter dropping to near zero again. The air supply can be regulated by a programmer to fit this daily pattern, preferably on a seven day cycle because weekend patterns will differ greatly from weekdays, but there must also be the ability to over-ride the programme on special occasions such as national or local holidays or sporting events creating increased traffic demand.

Manual over-ride must always be available, but whether or not in combination with automatic over-ride, there must be an alarm system bringing effectively to the notice of the controller abnormally high levels of CO polution or other dangerous conditions.

If circumstances permit ventilation and traffic flow control to be combined then economies in the maximum provision can be effected because the maximum flow of traffic through a tunnel does not make maximum demand on air. Maximum lane capacity is usually about 1750 vehicles when a speed of about 40 km/hr can be sustained. Vehicle spacing is then 23 m and pollutant output is less than half of that for the worst condition when traffic speed drops to 10 km/hr at a spacing of 10 m and flow to 1000 v.p.h.

A ventilation supply scheme regulated on the basis of the number of vehicles in the tunnel at any time, or preferably on measurement of average speed and spacing, appears to offer the best system for adquate and economical ventilation.

9.5.8 Air intake and discharge
The ideal situation for the intake is as close as possible to the point of supply to the tunnel in a position free from air pollution and protected from rain and snow. It is also important to avoid a site where noise either from the fans or from any intake screen is objectionable to neighbours.

Polluted air intake in urban conditions can arise from street traffic on the surface or from industrial processes in the vicinity, perhaps when the wind is from a particular direction. An example is at the Dartford Tunnel under the Thames where cement works adjacent to the river discharge dust into the atmosphere, which is drawn into the ventilation ducts when the wind is from one direction, and has to be cleaned out from time to time although otherwise harmless. The Tyne Tunnel likewise can suffer from adjacent industrial pollution.

Unless some settling chamber can be provided in the intake, or a sealed duct is provided, snow and fine rain can be drawn in by the fan and can affect the motors and control gear.

Noise from fans can be kept to a low level by good design, and particularly by keeping down blade tip speeds. If the environment is industrial there is little difficulty in complying with any reasonable noise standard, but in the vicinity of

Fig. 9.4 — Ventilation building for second Dartford Tunnel. Intake is through circular louvred chamber. Exhaust is through inverted conical evasee. Fans and equipment are on ground floor.

dwelling houses or hospitals it may be more difficult, and acoustic treatment may be necessary, with consequent added air resistance and cleaning problems. Noise is particularly objectionable and noticeable at night, but traffic is then normally very light and fan speeds can be minimal. It may even be possible to operate a tunnel at night on a varied circulation which dispenses with use of the fans in a sensitive area.

The problems of discharge are similar, but with the vital proviso that vitiated air should not be able to recirculate to the intake under any conditions. The ideal system in urban areas is probably discharge at high level through a tower or chimney acting as an evasée in the form of a cone whereby velocity head is transformed to suction head, contributing to economy of operation.

In the simpler systems where discharge is at a portal the questions of re-circulation and local environmental pollution likewise require careful study. One possibility in a shallow tunnel is to exhaust air by means of a number of jet fans straight up through the crown of the tunnel. This may be particularly economical where cut and cover construction is used and where the surface area is reclaimed as an open space. It must be remembered that the vitiated air is not usually at a dangerous level of pollution, that it is discharged vertically and will continue to rise being hot, and that in the open air there is rapid natural dilution.

9.5.9 Construction problems
Apart from the problems faced in all tunnelling work, ventilation has a few special aspects which must be considered at an early stage in the design. If a major tunnel is to have as few ventilation stations as practicable they will be large structures whose siting is closely determined by the alignment of the tunnel and the position of the portals. There will be heavy foundation loads to be carried down in the immediate vicinity of the tunnel and large shafts will be required as air ducts for both supply and exhaust.

The openings from these ducts into the tunnel present a special problem as large smooth continuous ducts without abrupt bends are desired. In any ground this presents difficult tunnelling conditions, and in bad ground aerodynamic efficiency may have to give way to practicalities of construction.

Construction of the duct connections and a short length of the tunnel in a caisson was adopted at the Dartford tunnel and elsewhere, the bored tunnel terminating by driving the shield into this chamber. At the Mersey Kingsway Tunnel the junctions were in rock and could be constructed in a series of headings but always adapted to the boring and lining of the main tunnel by machine.

9.6 VENTILATION IN FIRES
The frequency of vehicle fires in tunnels is not very different from that on the open road. A rate of one fire per km for every 10 million vehicles is representative.

The most important special feature is that the flames, smoke, and fumes cannot ascend vertically but rise to the roof of the tunnel and spread laterally, still burning and forming a layer possibly 2 or 3 m thick which travels horizontally at a rate of a few metres per second. Although there may be a zone of fresh air below the smoke cloud, heat radiated downwards may constitute a hazard to life. The spread and behaviour of this smoke cloud is greatly influenced by the ventilation air flow, and it is therefore important that fire fighters should be familiar with the possibilities of regulating air supply and extraction by selection of the most suitable combination of fans. Longitudinal ventilation without extraction presents a particular hazard of the spread of fire along the tunnel, but at least has the merit of supplying fresh air upstream from the fire and thus allowing access. Too high a rate of flow creates turbulence and the smoke obscures the whole tunnel. Semi-transverse ventilation has also a longitudinal flow, but its average velocity is usually lower and the continuous supply of fresh air at road level may be helpful. Fully transverse ventilation helps to localise the fire hazard, but the normal extraction rate may be quite inadequate to clear the smoke generated, and longitudinal spread in both directions may result, unless emergency outlets can be brought quickly into use.

In all types the hazard from a small fire, as from a single car, is not found to be severe; it is likely that the fire will remain localised, not spreading to adjacent vehicles, even at fairly close spacing. In a larger fire, the high temperature of the smoke cloud may make access to the seat of the fire difficult or dangerous. The problems of fire in a tanker carrying liquid fuel are such as to justify strict rules about the carrying of dangerous loads.

There is not complete agreement on the control of ventilation to the best advantage, but use of exhaust fans to extract smoke and to limit the spread of smoke to a single direction is certainly helpful, provided that the fans are not liable to be burnt out. The supply of fresh air is a help to fire fighters requiring access, but in major fires possible increased rate of combustion is advanced as a reason for restricting air supply. On the other hand, incomplete combustion tends to generate lethal quantities of carbon monoxide.

The particular requirements of each tunnel must be examined in the light of controls and facilities available and general policy on the hazards. In fact there have been very few major fires in highway tunnels. Two instances are described in Chapter 6.

9.7 RAILWAY VENTILATION

Ventilation is a problem of long tunnels. In a tunnel of moderate length the passage of a train creates a strong current of air so that polluted air is carried out to the exit portal while fresh air is drawn in behind.

Smoke was troublesome from the earliest days of railways. In addition to being environmentally objectionable to passengers, and to drivers, it produced corrosive deposits within the tunnel. Although dirt and unpleasantness in some degree were tolerated as inevitable, much was done to reduce smoke pollution by means of open shafts to the surface and by extract fans.

An early example of smoke extraction was on the world's first passenger railway at Liverpool terminus, when, after 40 years, in 1870 loco working was introduced in place of rope haulage on the 1 in 97 gradient from Edge Hill down to Lime Street. A 9 m diameter extract fan running at 45 r.p.m. was installed half way along the 1850 m tunnel. The tunnel was subsequently opened out as a cutting and widened to take four tracks. Another tunnel ventilation scheme was in the nearby Mersey Railway tunnel (1886) in which there were provided ventilation headings connecting to two extract fans in shafts on each side of the river crossing. The total output was 300 m^3/s, fresh air being drawn in through the stations. The Severn railway tunnel (1886) also had to be provided with ventilating fans before it was allowed to operate. These examples are not of course typical of the average railway tunnel: the conditions were particularly unfavourable to smoke clearance in that the Lime Street station is a terminus on the approach to which the trains were moving too slowly to be effective in sweeping out the polluted air, while the Mersey and Severn railway tunnels are subaqueous and deep with no possibility of open shafts.

In the first metro, the London Metropolitan Railway (1863), coal burning steam locos were employed, the last of which were not withdrawn until 1905. Ventilation of the shallow tunnels was by 'blow holes' of area about 90 m^2 at intervals of about 1 km. Smoke was a serious deterrent to passengers until electrification of the system.

The electrically operated London 'tube' railways which followed could not be provided with blow holes for ventilation and depended on piston action of trains which fitted closely into the twin tunnels, and exchange of air through station passages. Progressively, air supply to stations by fans has been found necessary to eliminate hot spots and the build up of heat. In the Victoria Line (1968-1971) reversible fans midway between stations were incorporated in the original design and are operated as exhaust fans, fresh air being drawn in through station entrances. Draught relief shafts are also provided at stations to suit the particular conditions. As a quantitative example, one length of 1 km incorporating one station and carrying 300 trains a day at a maximum speed of 18.6 m/s involves heat gain of 14.6 MWh, for which the extract fan provision is 40.5 m^3/s of air based on a temperature rise of 10.5°C, after allowing a 15% exchange of air through draught relief shafts.

In New York the main lines from the north into the Grand Central Station passed in a shallow tunnel 3 km long under Park Avenue. Steam operation gave rise to many difficulties and complaints: signals were obscured, passengers endured smoke and heat, and the public in the street above were subjected to blasts of ash, smoke and fumes through the ventilation openings at road level. Electrification was seen as the remedy, and after much careful study was adopted and effectively introduced in 1907.

Electrical operation has abolished the smoke problem for all metros and many main line railways, but other aspects of air control have become of growing importance. For tunnelled metro systems draught relief in stations and long term removal of accumulating heat are universal problems. For main line railways high operating speeds have introduced new problems of pressure pulses and train resistance more particularly in long tunnels.

The major differences from highway tunnel ventilation lie in:

(1) the absence of toxic gases,
(2) the self ventilating piston effect of trains, which occupy a substantial proportion of the tunnel cross section,
(3) the very high speeds attainable,
(4) the possibility of sealing trains against external atmospheric conditions,
(5) closely controlled use by stock of known characteristics on a fixed path.

Ventilation and aerodynamics cannot be entirely separated, but in the railway tunnel context ventilation may be considered to comprise the functions of removal of vitiated air and heat and, perhaps, air conditioning, while aerodynamics in its simplest aspect deals with the calculation and limitation of air speeds in public areas. Where operating speeds are high more complex problems arise, of air flows and pressures and transient pulses, which can be environmentally unacceptable, and also may call for substantial increases in loco operating power.

The impact of these problems differs considerably in metros and main line railways and may be considered for each type of system under the two aspects of environmental effects on passengers and staff and operational effects on plant and equipment. The control and adequacy of ventilation in the event of fire in a tunnel may be of particular importance in metros, but emergency provisions for long main line tunnels may also be thought necessary.

9.7.1 Metros

In tunnelled metros there are four zones used successively by passengers, namely,

(1) entrances and ticket halls
(2) escalators, stairs and access passages
(3) station platforms
(4) trains.

The environment in each is determined in part by the outside climatic conditions and in part by the ventilation system adopted. The general flow of air may either be inward from station entrances to platforms and tunnels, or outwards through escalator tunnels to discharge at station entrances. The former is generally preferable in ensuring fresh uncontaminated air in public areas, but in hot or cold extremes of climate or weather the unconditioned air entering stations may not be favoured.

In metros train speeds are not high because of the short runs between stations; the more complex aerodynamic problems do not therefore obtrude.

The ventilation system is required to ensure:

(1) an adequate supply of fresh air at acceptable temperatures,

(2) control of air flows induced by movement of trains where these create objectionable draughts in passenger areas,

(3) avoidance of abrupt and excessive temperature differences,

(4) removal of excess heat from all sources, which would otherwise accumulate within the metro.

Passage of a train through a tunnel generates considerable air movements; air will be expelled in front of a train and drawn in behind it through any ventilation shafts or other openings, including station passages. The effective exchange of stale air for fresh air will obviously depend on such factors as the proportion of tunnel cross-section occupied by the train, the area and length of shafts or other openings and their siting, the frequency of the train service and − to a major extent − on whether twin single track tunnels or double track two-way tunnels are used.

Self ventilation of the system by such means has generally been found to provide adequate quantities of fresh air, but fans are likely to be required to improve and regulate distribution, and more especially to deal with the removal of excess heat, which eventually is likely to become the major problem. The contingency of fire is also a reason for provision of fans, so designed and controlled as to allow smoke to be extracted in the most suitable path which leaves clear a passenger escape route. Possible heat damage to fans and their supply cables needs to be examined.

Having ensured the adequacy of the fresh air supply, it must be so controlled as to avoid draughts in passenger areas. The surges of air as trains enter and leave underground stations necessarily create air flows on platforms, in cross passages, and in escalator shafts and other connecting passages. These flows can be calculated from a knowledge of the layout of the tunnel system, including stations and of the operating characteristics of the trains − speed, acceleration, braking and frequency of service. Air speed of 7 m/s may be acceptable as a maximum, but 4 m/s is preferable. Excessive air flows can be reduced by such means as enlargement of passage areas, and, sometimes, provision of additional draught relief openings or passages in critical places, but, usually, the most effective action is the provision of draught relief shafts to the open air from the point of junction between running tunnels and station tunnels. Where the distance between stations so requires, ventilation shafts about half way between stations may also be provided.

The climate within a tunnel system will usually fluctuate less than outside, but with an inevitable tendency for the air to heat up, and for the structures also to become somewhat warmer than the long term outside average. In temperate climates there is not often any serious immediate problem, but at station entrances there may be some discomfort if air is drawn in when the weather is abnormally cold or hot.

In tropical climates refrigeration of the air supply may be thought necessary, but the problems of regulating temperatures in the various passage areas are very great.

Any attempt to cool the air in stations is seriously impaired by the exchange of air between stations and running tunnels arising from the piston action of trains. The construction of a separating wall between platform and train, with controlled openings to allow passengers to enter trains and alight, has been proposed and would effect substantial energy savings if found safe and practicable.

Air conditioning in trains is also at a disadvantage because of the opening of doors at frequent stations, and also because extra waste heat is generated by the refrigeration process and must be removed, probably by additional ventilating air to the tunnel. The air conditioning within a train has also to be adaptable to surface operation where trains run partly in a central tunnelled area and partly at the surface in outer suburban areas.

9.7.1.1 Heat gain

The accumulation of heat within a metro system is a phenomenon that may only become apparent over a period of years, but may appear quite quickly if traffic is heavy. The principal source of heat is the energy used in driving the trains, mostly converted to heat in the braking system but some also more directly in the driving motors. The major traction power requirement is for frequent and rapid acceleration out of stations: also significant are the lighting currents in trains and in stations and the body heat of large numbers of passengers. The first effect of these heating sources is to warm up the air in the system, which in turn transfers some of the heat to the tunnel lining and other structures and carries away some when it is exhausted.

The ground behind the lining acts as a 'heat sink' absorbing heat from the lining, or returning it if the air is unusually cold. This balancing effect evens out fluctuations between day and night and between summer and winter, but in the longer term average lining temperatures build up until there is no net annual loss of heat to the ground. The ventilating air has then to carry away the whole of the heat generated and its temperature rise is determined accordingly. It is for this condition that the ventilation system must ultimately be designed: the air flow must be sufficient in quantity to carry away the heat at the specified temperature rise, and must be directed so as to avoid discomfort to passengers. The environmental conditions in continuing hot weather are likely to determine maximum requirements. Conditions of severe cold have also to be allowed for.

Heated air can be extracted through fans in vent shafts at the end of stations or midway between stations. Another possibility is to site an extract duct under station platforms, which has the advantage of direct removal of much of the heat generated in the brake linings as the train slows down to enter the station. Any such extract system must draw in the same volume of air elsewhere, probably through station entrances.

In addition to the comfort and convenience of passengers and staff, ventilation must be related to the operational efficiency of plant and equipment. In general, electric motors and cables suffer loss of efficiency as their temperature rises, and therefore cooling by adequate ventilation is to be sought in the design of the system. Fans are less effective in handling warm air than cold and appreciable allowances must be made for extract fans. Fans must also be protected against surges of air or designed to withstand them.

9.7.1.2 Dust

A problem of some importance in metros is that of dust in tunnels. The dust is likely to contain metallic particles from the wear of wheels and rails and brake shoes. Dust is also brought into stations from the streets. Metallic dust blown about by passage of trains within the tunnels can be particularly deleterious to mechanical and electrical equipment, more especially to any delicate electrical contact devices, which must therefore be carefully protected against it. Removal of the dust by extract ventilation alone is not often practicable.

It will be clear that the requirements for ventilation of a metro system are varied and complex. In systems which have been built up progressively over many years, ventilation needs have become apparent concurrently with other development. Modifications and additions have been necessary in existing lines and the experience gained has been incorporated in new construction. In newer metros ventilation systems are almost invariably part of the initial design, but have not always proved adequate in operation. Until recent years calculations of sufficient accuracy for a complex system of tunnels have not been practicable, but computer programmes have now been devised and tested giving the necessary information for design.

9.7.1.3 Environment

Environmental standards required, if public transport services are to attract passengers, have risen in recent years and at the same time train frequency and speed have also risen so that the ventilation systems become more extensive and elaborate. It is all the more necessary that their planning and design is integrated with that of the system as a whole. The costs and benefits cannot be studied in isolation. All tunnelling operations for the provision of shafts, ducts, and underground plant rooms are necessarily difficult and costly, and wherever possible must be adapted and often subordinated to the tunnel construction as a whole. In urban areas the surface sites desirable for ventilation shafts and plant may be unobtainable. Shafts necessary for construction access may serve also as ventilation shafts, and, in improvement of ventilation in existing stations, old lift shafts may be utilised.

Environmental conditions have been examined in some detail by the U.S. Department of Transportation in their Subway Environmental Design Handbook Vol. I (1975) and limits of temperature, air flow speeds and air pressure pulses have been proposed. It is correctly suggested that increased attention to these factors has become necessary for the reasons given above, namely: greater heat loads for heavier faster and more frequent trains and higher environmental standards. It is questionable how far the pursuit of comfort should be carried when its capital and operating costs and their impact on fares are fully considered. The primary function of a metro system is to provide a fast and efficient service at a low cost. Cushioned armchair comfort and protection from every vagary of wind and weather can push costs up to excessive levels.

In a new system early and comprehensive studies are, of course, essential if appropriate provisions are to be incorporated in the system at the construction stage, having regard to the greatly enhanced cost and difficulty of effecting later improvements. Development of the ventilation design requires close liaison and interaction with those responsible for all the other aspects. In this, as in other aspects it may be just as important to reject unnecessarily high standards, after proper examination, as to provide for every possible contingency of discomfort.

9.8 MAIN LINE RAILWAYS

For main line railway tunnels, other than in the vicinity of termini, problems of fresh air supply arise only in long tunnels. Aerodynamic problems of pressure pulses, and train resistance are a consequence of high speed operation in tunnels of any length, becoming significant where speeds much exceed 120 km/h, and specially so in very long tunnels.

For main line tunnels less than a few kilometres long, self ventilation by train movement is usually fully adequate both for quality of air supply and elimination of heat build up. In twin single track tunnels the one way piston effect is particularly effective. In longer tunnels vent shafts are likely to provide the required changes of air if the tunnel is not at too great depth and if the necessary surface sites are available. On electrified lines air pollution is minimal: the worst conditions are with coal burning locos hauling heavy trains up a severe gradient.

A high speed train entering a tunnel portal acts abruptly on the nearly stationary air in the tunnel and sets up a complex set of aerodynamic effects. The air ahead of the train is compressed sharply and a shock wave is set up, travelling forward through the length of the tunnel at a speed close to that of sound. At the far portal this wave is reflected back as a pulse of rarefaction and is experienced as a sudden reduction of pressure when it reaches the train. Similar, but normally smaller, rarefaction pulses are generated at any discontinuities such as open shafts or cross passages.

By the mechanism of these pulses, the column of air in the tunnel is set in motion forwards and a stream of turbulent air flows back, through the annulus between train and tunnel lining, to the rear of the train, where suction is generated.

The difference of air pressures at front and rear of train constitutes a braking force, as also does the frictional drag of the turbulent air, whose motion relative to the train is rapid, but relative to the tunnel lining much slower.

In a short tunnel these braking effects are less than in a long tunnel, because, (1) the smaller mass of stationary air is more easily projected forward, (2) the turbulent flow and resultant drag are also less, (3) the forces act against the train for the shorter time during which the train is in the tunnel. The energy loss and consequent reduced speed are unlikely to be significant. The critical aerodynamic factor for operation of high speed trains through short tunnels is environmental, namely the development of pressure pulses which, if too great, can cause discomfort to the ears of passengers.

Research on this effect by British Rail has led to the adoption of criteria of acceptability. The unit adopted for pressure change in the pulses is 1 kN/m^2, which is approximately 1% of atmospheric pressure. It is considered that rapid changes should not be allowed to exceed 3 kN/m^2, but a limit of 0.5 kN/m^2 should be set for frequently repeated pulses, such as occur in a tunnel with numerous cross passages or shafts. It may be noted that the recommendation of the U.S. Department of Transportation Handbook is that the rate of change should not exceed $0.41 \text{ kN/m}^2\text{s}$.

The magnitude and importance of these effects may be exemplified by considering a twin single track tunnel 3 km long between portals without intermediate openings, and trains 350 m long running at 200 km/h. (See Fig. 9.5(a)). It will take 54 seconds for any point on the train to pass through the tunnel. A pressure rise of 4 kN/m^2 occurs during the first 7 s, dropping slightly as the tail of the train enters the tunnel and then after a further 7 s dropping sharply to a deficiency of 1 kN/m^2 when the wave of rarefaction has returned from the far portal. Thereafter the pressure fluctuates through a diminishing range. The drop of 5 kN/m^2 is not acceptable, and reduction has to be effected either by construction of intermediate shafts, or cross passages to the adjacent tunnel, or speed limitation to 150 km/h, or enlargement of the cross section of the tunnel.

The case of a double track tunnel with trains in both directions is more complex. The pressure pulse at entry is much less because of the greater area of tunnel, but maximum pressure changes occur as two trains meet and pass. Since the positions of maximum pressure change will be fixed by the meeting point of the trains, pressure relief by provision of vent shafts is unlikely to be practicable because very close spacing would be required to give relief at all times. If the surges are unacceptably high and if speeds cannot be reduced, a larger area of tunnel is called for.

Where a long tunnel is to carry high speed passenger trains the balance of advantage, aerodynamically, appears to be in favour of twin single track tunnels linked by cross passages. Even in the most favourable circumstances it is difficult to ensure the primary objects of ventilation: fresh air, minimum drag, adequate cooling, and limitation of pulse pressures. This was particularly apparent in the

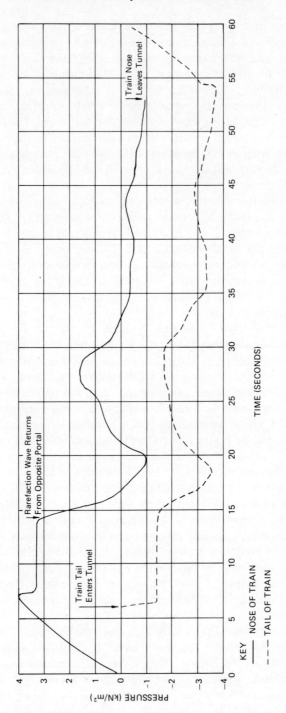

Fig. 9.5(a) — Pressure/time for train passing through single track tunnel (*Henson & Lowndes*).

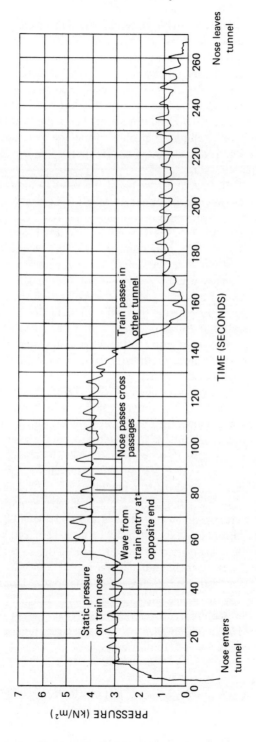

Fig. 9.5(b) – Typical static pressure on train nose in a twin bore system with cross passages (*Henson & Lowndes*).

1974 Channel tunnel scheme, where intermediate shafts were quite impracticable and a very frequent service of fast and heavy ferry trains was expected.

If the pressure and movement of air along a single track tunnel are to be kept within moderate limits, relief vents must be constructed, of sufficient cross sectional area. Where shafts are not practicable, cross passages to the adjacent tunnel serve the same purpose, except that they do not ensure exchange of air with the atmosphere and therefore do not discharge the heat building up in the system or relieve any pollution. In either form the vent openings constitute discontinuities reflecting back the pressure pulses, and as the train passes, generating new pulses.

The design of the size and spacing of the vents in terms of aerodynamics poses the choice between large openings widely spaced and smaller but more frequent openings. Draught relief is determined by vent area per unit length of tunnel but the magnitude of a pressure pulse is roughly proportional to the area of a single vent, and therefore more numerous smaller openings are advantageous in keeping pulses within specified limits. In the tunnel construction process numerous cross passages can be costly out of proportion to the direct cost because of disruption of the rate of advance of excavation and lining.

Figure 9.5b shows the pressure pulses for a train in a twin bore tunnel 10 km long with cross passages at 250 m intervals. A train at a speed of about 140 km/h enters the tunnel, and another enters at the far end 30 s later. The graph is of the static air pressure at the nose of the first train against time, the train taking 260 s to pass the length of the tunnel.

The pressure rises rapidly to about 3 kN/m^2 and after 50 secs is met by the pulse from the train in the other tunnel, rising to a little over 4 kN/m^2. As the train in the other tunnel passes, after 130 secs, the pressure drops quickly nearly to zero and only rises by about 1 kN/m^2 for the rest of the passage. Superimposed on these changes, each cross passage generates a sharp pulse of about 0.5 kN/m^2 at regular intervals of about 6½ secs. The instantaneous aerodynamic power follows a very similar curve rising to a maximum of 7 MW, falling back to 3 MW at the time of passing, and rising again to 6 MW. If relief vents were not provided much higher pressures would be encountered and there would be a severe penalty in power demand. From this example based on a speed of 140 km/h, the importance of minimising energy losses is obvious: even more so for higher speeds as pressures increase roughly as the square of the speed.

The significance of the energy demand and its control by proper design may be illustrated by figures from the Channel tunnel design of 1974 where the estimate of power required to overcome aerodynamic drag for a single ferry train was 13.5 MW in the absence of cross passages, reduced to 5.8 MW by provision of cross passages at 250 m spacing.

9.8.1 Heat gain

For short tunnels and moderate speeds self ventilation ensures that heat gain is not a serious problem, but as length of tunnel and operating speed increase, there may be an unacceptable rise in tunnel temperature. The operating power required to overcome aerodynamic drag increases as the square of the operating speed, and, once tunnel length increases beyond the capacity for self ventilation, all extra energy has to be otherwise exhausted. In the Channel tunnel the large heat input from the frequent train service could not readily be cooled by ventilation in the absence of shafts intermediate between those at the coasts, and therefore mechanical cooling systems were to be employed, fresh air being supplied normally through the service tunnel only to the extent of physiological demand, namely 50 m^3/s. In emergency, 200 m^3/s could be supplied through the interconnected servicing tunnels.

In the projected St. Gotthard Base tunnel proposed by Swiss Railways, the double track rail tunnel would be 48.67 km in length having a maximum gradient of 0.73% and with 3 deep shafts. Heat input from 282 trains a day and from fans and other plant is estimated at 25 MW, and in addition the high rock temperatures at depths of 1500 m would contribute 10 MW. Some 7 MW would be carried out by heating up of trains and half of the remainder by air supplied at a rate of 1600 m^3/s through two shafts and exhausted through the middle shaft and at the portals. It is proposed that the balance should be absorbed by vaporisation of water sprays.

BIBLIOGRAPHY

The subjects of ventilation and aerodynamics in vehicle tunnels are rapidly growing. As with compressed air working (Volume 1, Chapter 5), there are both physical and physiological aspects, the latter being a major factor in the ventilation of highway tunnels. Most of the information must be sought in papers presented at conferences and elsewhere, and in articles in the technical press. Many papers on the construction of individual vehicle tunnels, listed in the main bibliography, discuss their ventilation in some detail.

The following are considered of particular relevance. A very extensive bibliography, with abstracts, is contained in Pursall and King's book. The B.H.R.A. symposia provide comprehensive and up-to-date information.

Singstad, O., Ventilation of vehicular tunnels, *Proc. World Engng Congr., Tokyo*, 1929, **9**, 381-99.

Haldane, J. S., The ventilation of tunnels, *J. Instn Heat. Vent. Engrs*, 1936, **4** (Mar.).

Anderson, D., The Construction of the Mersey Tunnel, *J. Instn Civ. Engrs*, 1936, **2** (Apr.).

Murdock, C. W., Ventilating the Lincoln vehicular tunnel, *Trans. Amer. Soc. Heat. Vent. Engrs*, 1938, **44**, 273-88.

Daugherty, R. L., Piston effect of trains in tunnels, *Trans. Amer. Soc. Mech. Engrs*, 1942, **64** (Feb.).

Mount, S. C., Ventilation and cooling in London's tube railways, *J. Instn Heat. Vent. Engrs*, 1947, **14** (Jan.-Feb.).

Scheyer, E., Ventilation of Subways, *Mun. Engrs J.*, 1955, **41** (2).

Road Tunnels Committee:
Report, 11th World Congress, Rio de Janeiro, 1959;
Report, 12th World Congress, Rome, 1964;
Report, 13th World Congress, Tokyo, 1967;
Report, 14th World Congress, Prague, 1971;
Report, 15th World Congress, Mexico, 1975;
Report, 16th World Congress, Vienna, 1979;
Permanent International Association of Road Congresses, Paris.

Rennie, R. P. *et al.*, Effect of diesel locomotive operation on atmospheric conditions in a railway tunnel, *Can. J. Chem. Engng*, 1960, **38** (Aug.).

Ackeret, J. *et al.*, Ventilation of vehicle tunnels, *Mitt. Inst. Strassenbau, E.T.H. Zurich*, 1961 (10) (in German).

Brown, W. G., Temperature and ventilation in the Toronto subway, an experimental study, *Engr*, 1964, **218** (28 Aug.).

Brown, W. G., Basic theory of rapid-transit tunnel ventilation, *Trans. Amer. Soc. Mech. Engrs*, 1966, **88B** (Feb.).

Kennedy, E. R., *A mathematical model of tunnel ventilation*, Report 65-5, 1965.

Helly, W., *A mathematical model of tunnel ventilation*, Report 65-3, 1965.

Kennedy, E. R., *Port Authority tunnel ventilation: 1965-1966 experimental program*, Report 66-5, 1966.

Kennedy, E. R., *ditto, 1966-1967 experimental program*, Report 67-5, 1967.
Engineering Department, Port of New York Authority.

Fox, J. A. and Henson, D. A., The prediction of magnitudes of pressure transients generated by a train entering a single tunnel, *Proc. Instn Civ. Engrs*, 1971, **49** (May); *see also* discussion, 1971, **50** (Dec.).

The ventilation of road tunnels:
Constant, J., A general approach to the problem;
Cuaz, F., Mont Blanc tunnel;
Megaw, T. M., Road tunnels in U.K.;
Tunnels and Tunnelling, 1969, **4** (May and July).

The Aerodynamics and Ventilation of Vehicle Tunnels:
Proc. 1st Int. Symp., Canterbury, 1973;
Proc. 2nd Int. Symp., Cambridge, 1976;
Proc. 3rd Int. Symp., Sheffield, 1979.
British Hydromechanics Research Association.

Henson, D. A. and Fox, J. A., Transient flows in tunnel complexes of the type proposed for the Channel Tunnel, *Proc. Instn Mech. Engrs*, 1974, **188** (15/74).

Associated Engineers, *Subway Environmental Design Handbook*, U.S. Department of Transportation, 1975.

Pursall, B. R. and King, A. L., *The Aerodynamics and ventilation of vehicle tunnels: a state of the art review and bibliography*, British Hydromechanics Research Association, 1976.

Henson, D. A. and Lowndes, J. F. L., Design of a Ventilation system for an English Channel Tunnel, *ASHRAE J.*, 1978 (Feb.).

Bibliography

This is confined to publications in the English language, and is classified in six sections, the items in each section being listed in order of date of publication:
A. Books and reports
B. Institution of Civil Engineers
C. American Society of Civil Engineers
D. Other organisations
E. Conferences and Symposia
F. Journals and Periodicals
Further extensive bibliographies can be found in Pequignot (1963), Pursall and King (1976) and Hoek and Brown (1980) as listed in Section A.

A. BOOKS AND REPORTS

Simms, F. W., *Practical tunnelling*, Crosby Lockwood, 1844 (revised and enlarged, 1859, 1877, 1896).

Brunel, I., *The life of Isambard Kingdom Brunel, Civil Engineer*, Longmans, 1870.

Drinker, H. S., *Tunneling, explosive compounds and rock drills*, J. Wiley, 1878.

Gripper, C. F., *Railway tunnelling in heavy ground*, Spon, 1879.

Burr, S. D. V., *Tunnelling under the Hudson River*, J. Wiley, 1885 (reprinted A. A. Mathews, Arcadia, 1968).

Walker, T. A., *The Severn Tunnel*, Bentley, 1888 (reprinted Kingsmead, 1969).

Copperthwaite, W. C., *Tunnel shields and the use of compressed air in sub-aqueous work*, Constable, 1906 (reprinted A. A. Mathews, Arcadia, 1968).

Gilbert, G. H. and others, *The Subways and tunnels of New York*, J. Wiley, 1912.

Hewett, B. H. M. and Johannesson, S., *Shield and compressed air tunneling*, McGraw Hill, 1922 (reprinted A. A. Mathews, Arcadia, 1968).

The McAlpine system of reinforced concrete tunnel lining, Sir Robert McAlpine & Sons, 1935.

Noble, C., *The Brunels*, Cobden-Sanderson, 1938.

Richardson, H. W. and Mayo, R. S., *Practical tunnel driving*, McGraw-Hill, 1941 (revised edn., R. S. Mayo, 1975).

Terzaghi, K., Rock defects and loads on tunnel supports: Introduction to tunnel geology, in Proctor, R. V. and White, T. L., *Rock tunneling with steel supports*, Commercial Shearing and Stamping Co., Ohio, 1946, 2nd edn., 1968.

Terzaghi, K., Geologic aspects of soft-ground tunneling, Ch. 11 in Trask, P. D., (ed), *Applied Sedimentation*, J. Wiley, 1950.

Rolt, L. T. C., *Isambard Kingdom Brunel*, Longmans, 1957.

Slater, H. and Barnett, C., *The Channel Tunnel*, Wingate, 1958.

The Work in Compressed Air Special Regulations, SI 1958 no. 61, HMSO, 1958.

Pratt, A. W. and Daws, L. F., Heat transfer in deep underground tunnels, *National Building Studies Research Paper 26*, HMSO, 1958.

The Stockholm Underground Railway: a technical description 1958 and *A technical description of the Stockholm Underground 1964*, Public Works Dept. and Passenger Transport Co., Stockholm.

Payne, R., *The canal builders*, Macmillan, New York, 1959.

Hammond, R., *Tunnel engineering*, Heywood, 1959.

Jeffery, A. H. G., *Immersed tube tunnels*, Kemp's, 1962.

Jackson, A. A. and Croome, D. F., *Rails through the clay: a history of London's tube railways*, Allen & Unwin, 1962.

Lampe, D., *The Tunnel: the story of the world's first tunnel under a navigable river, dug beneath the Thames 1824–42*, Harrap, 1963.

Ministry of Transport, *Proposals for a fixed Channel link, Cmnd 2137*, HMSO, 1963.

Pequignot, C. A. (ed.), *Tunnels and tunnelling*, Hutchinson, 1963.

Sandström, G. E., *The history of tunnelling*, Barrie & Rockliff, 1963.

Blower, A., *British railway tunnels*, Ian Allan, 1964.

Guthrie Brown, J., *Hydroelectric engineering practice, 1 Civil Engineering*, Blackie, 1964.

Howson, F. H., *World's underground railways*, Ian Allan, 1964.

Reid, A. S. H., Road tunnels more than 500 feet long, *Road Research Technical Paper 78*, HMSO, 1965.

Coleman, T., *The railway navvies*, Hutchinson, 1965.

Pequignot, C. A., *Chunnel*, C. R. Books, 1965.

Havers, H. C. P., *Underground railways of the world*, Temple Press, 1966.

Terzaghi, K. and Peck, R. B., *Soil mechanics in engineering practice*, J. Wiley, 2nd edn., 1967.

McGregor, K., *The drilling of rock*, C. R. Books, 1967.

Proctor, R. V. and White, T. L., *Rock tunneling with steel supports* (incl. Terzaghi, q.v.), Commercial Shearing & Stamping Co., Ohio, 2nd edn., 1968.

Stagg, K. G. and Zienkiewicz, O. C., *Rock mechanics in engineering practice*, J. Wiley, 1968.

Follenfant, H. G., *Underground railway construction*, London Transport, 1968.

Mayo, R. S. and others, *Tunneling: the state of the art*, U.S. Department of Housing and Urban Development, (NTIS PB 178036), 1968.

Safety in sewers and at sewage works, Institution of Civil Engineers, 1969.

Deere, D. U. and others, *Design of tunnel liners and support systems*, U.S. Department of Transportation (NTIS PB 183799), 1969.

Peck, R. B. and others, *Some design considerations in the selection of underground support systems*, U.S. Department of Transportation (NTIS PB 190443), 1969.

Household, H., *The Thames and Severn Canal*, David & Charles, 1969.

Clements, P., *Mark Isambard Brunel*, Longmans, 1970.

Duncan, J. M. and others, *Materials handling for tunneling*, U.S. Department of Transportation Report FRA-RT-71-57 (NTIS PB 197331), 1970.

Fenix and Scisson Inc., *A systems study of soft-ground tunneling*, U.S. Department of Transportation Report DOT-FRA-OHSGT-231 (NTIS PB 194769), 1970.

Major Tunnels in Japan, 1970, Japan Society of Civil Engineers.

Howson, F. H., *The rapid transit railways of the world*, Allen & Unwin, 1971.

Parker, H. W. and others, *Innovations in tunnel support systems*, U.S. Department of Transportation Report FRA-RT-72-17 (NTIS PB 204437), 1971.

Beaver, P., *A History of tunnels*, P. Davies, 1972.

Safety in wells and boreholes, Institution of Civil Engineers, 1972.

I.C.I., *Blasting Practice*, Nobel's Explosives Co. Ltd., Scotland, 4th edn., 1972.

Underground Construction Research Council, *The use of underground space to achieve national goals*, American Society of Civil Engineers, 1972.

Department of the Environment, *The Channel Tunnel Project, Cmnd 5256*, HMSO, 1973.

Department of the Environment, *The Channel Tunnel, Cmnd 5430*, HMSO, 1973.

Haining, P., *Eurotunnel, an illustrated history of the Channel scheme*, New English Library, 1973.

Nock, C. S., *Underground railways of the world*, A. & C. Block, 1973.

Parker, H. W. and others, *Testing and evaluation of prototype tunnel support systems*, U.S. Department of Transportation Report FRA-ORDD 74-11 (NTIS PB 231912), 1973.

Paul, S. L. and others, *Research to improve tunnel support systems*, U.S. Department of Transportation Report FRA-ORDD 74-51 (NTIS PB 235762), 1973.

Semple, R. A. and others, *The effect of time-dependent properties of altered rock on tunnel support requirements*, U.S. Department of Transportation Report FRA-ORDD-74-30 (NTIS PB 230 307), 1973.

Sverdrup & Parcel and Associates, *Cut-and-cover tunneling techniques, 1, A study of the state of the art, 2, Appendix*, U.S. Department of Transportation Reports FWHA-RD-73-40 & 41 (NTIS PB 222997 & 8), 1973.

Szechy, K., *The art of tunnelling*, Akademiai Kiado, 2nd edn., 1973.

Wahlstrom, E. E., *Tunneling in rock*, Elsevier, 1973.

Blyth, F. G. H. and de Freitas, M. H., *A geology for engineers*, E. Arnold, 6th edn., 1974.

Jacobs Associates, *Ground support prediction model (RSR concept)*, U.S. Bureau of Mines (NTIS AD 773018), 1974.

Follenfant, H. G., *Reconstructing London's underground*, London Transport, 2nd edn., 1975.

Cording, E. J. and others, *Methods for geotechnical observations and instrumentation in tunneling*, 2 vols., Dept. of Civ. Engng, University of Illinois, 1975 (NTIS PB 252585 & 6).

The present status of tunnelling activities in Japan, Japan Tunnelling Association, 1975.

Clough, G. W., *A Report on the practice of chemical stabilisation around soft ground tunnels in England and Europe*, Dept. of Civ. Engng, Stamford University, 1975.

Associated Engineers, *Subway Environmental Design Handbook*, U.S. Department of Transportation, 1975.

Muir Wood, A. M., 'Tunnelling', ch. 30 in Blake, L. S. (ed.), *Civil Engineer's Reference Book*, Newnes-Butterworth, 3rd edn., 1975.

Bell, F. C. (ed.), *Methods of treatment of unstable ground*, Butterworths, 1975.

Cording, E. J., *Displacements around tunnels in soil*, U.S. Department of Transportation Report DOT-TST 76T-22 (NTIS PB 267356), 1976.

Glerum, A. and others, Motorway tunnels built by the immersed tube method, *Rijkswaterstaat Communications*, 1976 (25).

Manual of applied geology for engineers, Institution of Civil Engineers and HMSO, 1976.

Herndon, J. and Lenahan, J., *Grouting in Soils*, 2 vols., U.S. Federal Highway Administration Reports FHWA-RD-76-26 & 27 (NTIS PB 259043 & 4), 1976.

Harding, H. J. B., Tunnels, ch. 2 in Pugsley, A. (ed.), *The works of Isambard Kingdom Brunel*, Institution of Civil Engineers, 1976, and Cambridge U.P., 1980.

Golder Associates and James F. MacLaren Ltd., *Tunnelling technology*, Ontario Ministry of Transport and Communications, 1976.

Pursall, B. R. and King, A. L., *The aerodynamics and ventilation of vehicle tunnels: a state of the art review and bibliography*, British Hydromechanics Research Association, 1976.

Parsons, Brinckerhoff, Quade & Douglas, *Subsurface exploration methods for soft ground rapid transit tunnels*, 2 vols., U.S. Department of Transportation Report UMTA-MA-06-0025-76-1 & 2 (NTIS PB 258343 & 4), 1976.

Bechtel Inc., *Systems study of precast concrete tunnel liners*, U.S. Department of Transportation Report DOT-TSC-OST-77-7 (NTIS PB 264761), 1977.

Consulting Engineers Group, *Prefabricated structural members for cut and cover tunnels, 1, Design concepts, 2, Three case studies,* U.S. Department of Transportation Reports FHWA-RD-76-113 & 4 (NTIS PB 273530 & 1), 1977.

DeLeuw, Cather & Co., *Study of subway station design and construction,* U.S. Department of Transportation Report UMTA-MA-06-0025-77-6 (NTIS PB 268894), 1977.

Proctor, R. V. and White, T. L., *Earth tunneling with steel supports,* Commercial Shearing & Stamping Co., Ohio, 1977.

Brakel, J., *Submerged Tunnelling,* Technische Hogeschool, Delft, 1978.

BS 5573 Code of Practice for safety precautions in the construction of large diameter boreholes for piling and other purposes, (Revision of CP 2011), British Standards Institution, 1978.

Langefors, U. and Kihlstrom, B., *The modern technique of rock blasting,* J. Wiley, 3rd edn., 1978.

Bayliss, D. A., *The Post Office Railway London,* Turntable Pubns., 1978.

Ranken, R. E. and others, *Analysis of ground-liner interaction for tunnels,* U.S. Department of Transportation Report UMTA-IL-06-0043-78-3 (NTIS PB 294818), 1978.

Metropolitan Atlanta Rapid Transit Authority, *The Atlanta research chamber; applied research for tunnels: blasting techniques, conventional shotcrete, steel-fiber-reinforced shotcrete; monographs on the state-of-the-art of tunneling,* U.S. Department of Transportation Report UMTA-GA-06-0007-79-1 (NTIS PB 297574), 1979.

MacPherson, H. H. and others, *Settlements around tunnels in soil: three case histories,* U.S. Department of Transportation Report UMTA-IL-06-0043-78-1 (NTIS PB 290856), 1978.

O'Rourke, T. D., *Tunneling for urban transportation: a review of European construction practice,* U.S. Department of Transportation Report UMTA-IL-06-0041-78-1, 1978.

Hadfield, C., *British canals,* David & Charles, 6th edn., 1979.

Jaeger, C., *Rock mechanics and engineering,* Cambridge U.P., 2nd edn., 1979.

Schach, R. and others, *Rock bolting, a practical handbook,* Pergamon, 1979.

Rogers, L. H. (ed.), *International statistical handbook of urban public transport,* International Union of Public Transport (U.I.T.P.), Brussels, 1979.

Hoek, E. and Brown, E. T., *Underground excavations in rock,* Institution of Mining and Metallurgy, 1980.

The Present Status of Tunnelling Activities in Japan, Japan Tunnelling Association, Tokyo, 1980.

Schlaug, R. N., *Management of air quality in and near highway tunnels,* U.S. Department of Transportation Report FHWA-RD-78-186, 1980.

BS 5930, Code of Practice for site investigations (Revision of CP 2001), British Standards Institution, 1981.

Bowen, R., *Grouting in engineering practice*, Applied Science, 2nd edn., 1981.

House of Commons, Second Report from the Transport Committee, Session 1980-81, *The Channel Link, Report & Minutes, (HoC 155-I, II & III)*, HMSO, 1981.

Harding, H. J. B., *Tunnelling history and my own involvement*, Golder Assocs., 1981.

A guide to pipe jacking design, Pipe Jacking Association, 1981.

Stack, B., *Handbook of mining and tunnelling equipment*, J. Wiley, 1981.

BS 6164, Code of Practice for Safety in Tunnelling in the construction industry, British Standards Institution, 1982.

B. INSTITUTION OF CIVIL ENGINEERS

The principal publications embodying papers and discussions are:

Minutes of Proceedings	1837-1935,	**1-240**	MP
Journal	1935-1951,	**1-36**	J
Proceedings	1952- ,	**1-**	P

From Proceedings 1957, **8** (Sept) discussions were published later and are separately noted.

Additional series of papers are indicated:

Selected Engineering Papers	1922-1935	SEP	
Engineering Division Papers	1941-1951:	Works	Wks
		Railway	Rly
Civil Engineer in War	1948	CEW	
Chartered Civil Engineer	1949-1957	CCE	

'Description of the works on the Netherton tunnel branch of the Birmingham canal',

Walker, J. R., *MP*, 1860, **19**

'On the construction and enlargement of the Lindal tunnel, on the Furness railway',

Stileman, F. C., *MP*, 1860, **19**

'Description of the Lydgate and of the Buckhorn Weston railway tunnels',

Fraser, J. G., *MP*, 1863, **22**

'The actual state of the works on the Mont Cenis tunnel, and description of the machinery employed',

Sopwith, T., *MP*, 1864, **23**

'On the main drainage of London, and the interception of the sewage from the River Thames',

Bazalgette, J. W., *MP*, 1865, **24**

'On the geological conditions affecting the construction of a tunnel between England and France',

Prestwich, J., *MP*, 1873, **37**

'The St. Gothard tunnel',

Clark, D. K., *MP*, 1875, **42**

'On tunnel construction, and on the Sydenham tunnel, London, Chatham and Dover railway',
  Baldwin, A. E., *MP*, 1877, **49**
'The St. Gothard tunnel',
  Clark, D. K., *MP*, 1879, **57**
'Tunnel outlets from storage reservoirs',
  Wood, C. J., *MP*, 1879, **59**
'The main drainage of Torquay',
  Chatterton, G., *MP*, 1880, **61**
'Summit-level tunnel of the Bettws and Festiniog railway',
  Smith, W., *MP*, 1883, **73**
'The Metropolitan and Metropolitan District Railways',
  Baker, B.,
'The City lines and extensions (Inner Circle completion) of the Metropolitan and District railways',
  Barry, J. W., *MP*, 1885, **81**
'The Mersey Railway',
  Fox, F., *MP*, 1886, **86**
'Yanegase Yama tunnel on the Tsuruga-Nagahama railway, Japan',
  Hasegawa, K., *MP*, 1887, **90**
'Covered way as constructed on the Glasgow City and District railway',
  Wilson, W. S., *MP*, 1888, **92**
'Reparation of Betchworth tunnel, Dorking, on the London, Brighton and South Coast railway',
  Lopes, G., *MP*, 1888, **95**
'Alpine engineering',
  Vernon-Harcourt, L. F., *MP*, 1888, **95**
'Methods adopted in constructing the Glasgow Central railway (Bridgeton and Trongate contracts)',
  Barker, C. D., *MP*, 1893, **114**
'Method of underpinning and tunnelling under tenements',
  Young, J., *MP*, 1893, **115**
'Tunnels on the Dore and Chinley railway',
  Rickard, P., *MP*, 1894, **116**
'Timbering in the Ampthill second tunnel',
  Matheson, E. E., *MP*, 1895, **120**
'Glasgow District subway',
  Stewart, A. M., *MP*, 1895, **122**
'The City and South London Railway; with some remarks upon subaqueous tunnelling by shield and compressed air',
  Greathead, J. H., *MP*, 1895, **123**
'Bold Street extension tunnel and Central low-level station of the Mersey railway',
  Rowlandson, C. A., *MP*, 1895, **123**

'Enlargement of the City and South London Railway tunnels',
    Jones, I. J. and Curry, G., *MP*, 1927, **224**
'Sub-aqueous tunnelling in compressed air, with reference to Barking power-station cable tunnel under the River Thames',
    Matheson, D. S., *SEP*, 1927, (43)
'The reconstruction of Mossgiel tunnel',
    McLellan, D., *SEP*, 1927, (53)
'The new Piccadilly Circus station',
    Hall, H., *MP*, 1929, **228**
'The Tata Company's hydro-electric project',
    Cursetjee, N. J. M., *MP*, 1929, **228**
'The Lochaber water-power scheme',
    Halcrow, W. T., *MP*, 1930, **231**
'The opening-out of Cofton tunnel, London Midland and Scottish railway',
    McCallum, R. T., *MP*, 1930, **231**
'The construction, testing and strengthening of a pressure tunnel for the water supply of Sydney, N.S.W.',
    Haskins, G., *MP*, 1932, **234**
'Cementation in the Severn Tunnel',
    Carpmael, R., *MP*, 1932, **234**
'The River Hooghly Tunnel',
    Norrie, C. M., *MP*, 1933, **235**
'Power-house foundations and circulating-water tunnels at the Ford Motor Company's works, Dagenham',
    Allin, R. V. and Nachshen, M., *MP*, 1933, **235**
'Circulating-water culverts for the Portishead generating station of the Bristol corporation',
    Rees, G. M. G., *SEP*, 1934, (163)
'The coaling-jetty, circulating water system and cable-tunnels of the Battersea power station of the London Power Co. Ltd.',
    Bartlett, V. F. and Cadwell, W. H., *MP*, 1935, **240**
'The construction of the Mersey Tunnel',
    Anderson, D., *J.*, 1936, **2** (Apr.)
'The construction of infiltration galleries for the water-supply of Ndola, Northern Rhodesia',
    McIlleron, R. C., *J.*, 1936, **3** (June)
'West Middlesex Main Drainage',
    Watson, D. M., *J.*, 1937, **5** (Apr.)
'Reconstruction of Aldgate East station',
    Harley-Mason, J. H., *J.*, 1939, **11** (Apr.)
'The strengthening and final testing of the pressure tunnel for the water supply of Sydney, N.S.W.',
    Farnsworth, S. T., *J.*, 1939, **11** (Apr.)

'Notes on the station-lengthening at Shepherd's Bush station',
   Filor, C. G. H., *J.*, 1940, **13** (Feb.)
'Tunnel linings with special reference to a new form of reinforced concrete lining',
   Groves, G. L., *J.*, 1943, **20** (Mar.)
'The Laghouat escape tunnel',
   Brickell, R. G., *J.*, 1943, **20** (Apr.)
'Tunnelling with pre-cast concrete',
   Pattenden, B., *J.*, 1945, **23** (Feb.)
'Emergency repair to the Tower subway, London, after air-raid damage',
   Harding, H. J. B., *J.*, 1945, **25** (Nov.)
'Particle size in silts and sands',
   Glossop, R. and Skempton, A. W., *J.*, 1945, **25** (Dec.)
'The use of aluminous cement in the construction of the Mosul tunnel, Iraqi state railways',
   Hagger, G. C., *J.*, 1945, **25** (Dec.)
'The Ilford Tube',
   Groves, G. L., *J.*, 1946, **26** (Mar.)
'Some uses of explosives in civil engineering',
   Lorimer, J., *J.*, 1946, **26** (Mar.)
'Applications of the freezing process to civil engineering works',
   Mussche, H. E. and Waddington, J. C., *Wks*, 1946 (5)
'The choice of expedients in civil engineering construction',
   Harding, H. J. B., *Wks*, 1946 (6)
'Tunnel maintenance',
   Robarts, H. E., *Rly*, 1947 (23)
'The construction of the Trans-Iranian Railway, with particular reference to the works of "Lot 6 (South)" ',
   Champion, C. L., *J.*, 1947, **29** (Dec.)
'Some special storages',
   Temple-Richards, H. B., *CEW*, 1948, **3**
'Protection against flooding in the London Underground railway system',
   Morgan, H. D., *CEW*, 1948, **3**
'The tunnels in Gibraltar',
   Cotton, J. C., *CEW*, 1948, **3**
'Site investigations, including boring and other methods of subsurface exploration',
   Harding, H. J. B., *J.*, 1949, **32** (Apr.)
'Compressed-air caisson foundations',
   Wilson, W. S. and Sully, F. W., *Wks*, 1949 (13)
'The driving and lining of the Clunie tunnel on the Tummel-Garry hydro-electric project',
   Grundy, C. F., *Wks*, 1950 (15)

'The Mullardoch-Fasnakyle-Affric tunnels',
    Dillon, E. C., *Wks*, 1950 (16)
'Arley tunnel: remedial works following subsidence',
    King, C. W., *J.*, 1951, **36** (Mar.)
'Part reconstruction of Bo-Peep tunnel at St. Leonards-on-Sea',
    Campion, F. E., *J.*, 1951, **36** (Mar.)
'A 75-inch diameter water main in tunnel: a new method of tunnelling in London clay',
    Scott, P. A., *P.*, 1952, **1** (May)
'Civil engineering aspects of hydro-electric development in Scotland',
    Fulton, A. A., *P.*, 1952, **1** (July)
'Tunnelling plant and equipment',
    Archer, G. C., *P.*, 1952, **1** (Sep.)
'Some major problems in railway civil engineering maintenance',
    Cantrell, A. H., *P.*, 1953, **2** (Feb.)
'Special features of the Affric hydro-electric scheme (Scotland)',
    Roberts, C. M., *P.*, 1953, **2** (Sep.)
'The trend of development in the design and equipment of underground railways',
    Coombs, D. H. and Wilson, G. J., *P.*, 1953, **2** (Oct.)
'The construction of Tigres dam and Malgovert tunnel',
    Pelletier, J., *P.*, 1953, **2** (Dec.)
'The construction of two concrete-lined tunnels for the Machkund hydro-electric project, South India',
    Gauld, G. A. and Newport, R. J., *P.*, 1954, **3** (Mar.)
'Concerning the main drainage of London',
    Warlow, W. P., *P.*, 1954, **3** (May)
'Swedish underground hydro-electric power stations',
    Hagrup, J. F., *P.*, 1954, **3** (Aug.)
'Woodhead New Tunnel: construction of three-mile main double-line railway tunnel',
    Scott, P. A. and Campbell, J. I., *P.*, 1954, **3** (Sep.)
'Present trends in design of pressure tunnels and shafts for underground hydro-electric power stations',
    Jaeger, C., *P.*, 1955, **4** (Mar.)
'Investigations into the design of pressure tunnels in London clay',
    Tattersall, F. and others, *P.*, 1955, **4** (July)
'The diversion of the Annalong River into the Silent Valley reservoir',
    Colebrook, C. F., *P.*, 1956, **5** (Apr.)
'Uses of aerated cement grout and mortar in stabilization of slips in embankments, large scale tunnel repairs and other works',
    Purbrick, M. C. and Ayres, D. J., *P.*, 1956, **5** (Feb.)
'The construction of Middleton connecting sewer in north Manchester',
    Collier, E. H., *P.*, 1956, **5** (Dec.)

'The equipment of the Ilford tube',
 Cuthbert, E. W., *Rly*, 1950 (36)
'Pretreatment of gravel for compressed air tunnelling under the River Thames at Dartford',
 Kell, J., *CCE*, 1957 (Mar.)
'The introduction of tunnelling for the construction of large-diameter sewers in Kingston-upon-Hull',
 Roberts, G. W. and Lucas, W. H., *P.*, 1957, **6** (Apr.)
'Shaft raising in the Scottish Highlands',
 Henderson, R. S., *P.*, 1957, **6** (Apr.)
'Portishead "B" power station, with particular reference to the circulating-water works',
 Morgan, H. D. and others, *P.*, 1957, **7** (Aug.)
'Preliminary planning for a new tube railway across London',
 Turner, F. S. P., *P.*, 1959, **12** (Jan.), disc. 1959, **13** (Aug.)
'The use of the freezing process in the construction of a pumping station and storm-water overflow at Fleetwood, Lancs.',
 Ellis, D. R. and McConnell, J., *P.*, 1959, **12** (Feb.), disc. 1959, **13** (July)
'Underground station for Western District Post Office, London: design and control',
 Blackford, S. and Cuthbert, E. W.,
'Ditto: construction',
 Collingridge, V. H. and Tuckwell, R. E., *P.*, 1960, **15** (Feb.), disc. 1960, **17** (Nov.)
'Zambezi hydro-electric development at Kariba: first stage',
 Anderson, D. and others, *P.*, 1960, **17** (Sep.), disc. 1962, **21** (Mar.)
'The work of the Channel Tunnel Study Group, 1958–60',
 Bruckshaw, J. M. and others, *P.*, 1961, **18** (Feb.), disc. 1962, **21** (Mar.)
'New tunnels near Potters Bar in the Eastern Region of British Railways',
 Terris, A. K. and Morgan, H. D., *P.*, 1961, **18** (Apr.), disc. 1962, **21** (Apr.)
'The Thames to Lee tunnel water main',
 Cuthbert, E. W. and Wood, F., *P.*, 1962, **21** (Feb.), disc. 1962, **23** (Dec.)
'An introduction to alluvial grouting',
 Ischy, E. and Glossop, R., *P.*, 1962, **21** (Mar.), disc. 1962, **23** (Dec.)
'Special features of the civil engineering works at Aberthaw power station',
 Fraenkel, P. M. and Triggs, R. L., *P.*, 1962, **23** (Sep.), disc. 1964, **27** (Mar.)
'Installation of a reinforced-concrete invert in the Gillingham (Dorset) tunnel, on the Southern Region of British Railways',
 Tanner, P. W. and Burton, R. A. H., *P.*, 1963, **24** (Jan.), disc. 1964, **27** (Jan.)
'The subsidence at Fylde Street, Farnworth',
 Hale, H. T. and Dyer, E. A., *P.*, 1963, **24** (Feb.), disc. 1963, **26** (Sep.)
'The Dartford tunnel',
 Kell, J., *P.*, 1963, **24** (Mar.), disc. 1963, **26** (Dec.)

'Cameron Highlands hydroelectric scheme',
    Dickinson, J. C. and Gerrard, R. T., *P.*, 1963, **26** (Nov.), disc. 1964, **29** (Sep.)
'Park Lane improvement scheme: design and construction',
    Granter, E., *P.*, 1964, **29** (Oct.), disc. 1966, **33** (Mar.) and 1966, **35** (Dec.)
'Clyde tunnel: design, construction and tunnel services',
    Morgan, H. D. and others,
'Ditto: constructional problems',
    Haxton, A. F. and Whyte, H. E., *P.*, 1965, **30** (Feb.), disc. 1967, **37** (July)
'The Bretby tunnelling machine',
    Hay, J. D. and others, *P.*, 1965, **30** (Apr.), disc. 1966, **35** (Sep.)
'Victoria Line: experimentation, design, programming and early progress',
    Dunton, C. E. and others, *P.*, 1965, **31** (May), disc. 1966, **34** (July)
'Hazard from blasting fumes in normal rock tunnelling, with particular reference to nitrogen dioxide',
    Fogden, C. A. and Garrod, A. D., *P.*, 1965, **31** (July), disc. 1966, **34** (Aug.)
'Soft ground tunnelling for the Toronto subway',
    Bartlett, J. V. and others, *P.*, 1965, **32** (Sep.), disc. 1966, **34** (May)
'Blackwall tunnel duplication',
    Kell, J. and Ridley, G., *P.*, 1966, **35** (Oct.), disc. 1967, **37** (July)
'Strand underpass',
    Finn, E. V. and others, *P.*, 1966, **35** (Nov.), disc. 1967, **37** (Aug.)
'Some design and construction features of the Cruachan pumped storage project',
    Young, W. and Falkiner, R. H., *P.*, 1966, **35** (Nov.), disc. 1967, **37** (Aug.)
'Furnas hydro-electric scheme, Brazil: closure of diversion tunnels',
    Lyra, F. H. and MacGregor, W., *P.*, 1967, **36** (Jan.), disc. 1967, **37** (Aug.)
'Horizontal earth boring',
    Thomson, J. C., *P.*, 1967, **36** (Apr.)
'The Bristol regional foul water drainage scheme',
    Steel, P. H., *P.*, 1967, **38** (Oct.), disc. 1968, **41** (Oct.)
'The Tyne tunnel: planning of the scheme',
    Prosser, J. R. and Grant, P. A. St. C.,
'Ditto: construction of the main tunnel',
    Falkiner, R. H. and Tough, S. G., *P.*, 1968, **39** (Feb.), disc. 1968, **41** (Nov.)
'Cooling water intakes at Wylfa power station',
    Chapman, E. J. K. and others, *P.*, 1969, **42** (Feb.), disc. 1969, **44** (Nov.)
'Thames Cable tunnel',
    Haswell, C. K., *P.*, 1969, **44** (Dec.), disc. 1970, **47** (Oct.)
'The Victoria Line',
    Follenfant, H. G. and others, *P.*, 1969, **Suppl**, disc. 1970, **Suppl**
'The site investigations for a Channel tunnel 1964–65',
    Grange, A. and Muir Wood, A. M., *P.*, 1970, **45** (Jan.), disc. 1970, **47** (Dec.)

'Chesapeake Bay bridge-tunnel project',
  Sverdrup, L. J., *P.*, 1970, **46** (May)
'Construction of the Monmouth tunnels in soft rock',
  Parry, Rj. R. and Thornton, G. D., *P.*, 1970, **47** (Sep.), disc. 1971, **49** (July)
'Design and construction of the cargo tunnel at Heathrow airport, London',
  Muir Wood, A. M. and Gibb, F. R., *P.*, 1971, **48** (Jan.), disc. 1971, **50** (Oct.)
'Some considerations of submerged tunnelling',
  Brakel, J., *P.*, 1971, **48** (Apr.), disc. 1972, **51** (Jan.)
'The prediction of the magnitudes of pressure transients generated by a train entering a single tunnel',
  Fox, J. A. and Henson, D. A., *P.*, 1971, **49** (May), disc. 1971, **50** (Dec.)
'The Manapouri power project, New Zealand',
  Langbein, J. A., *P.*, 1971, **50** (Nov.), disc. 1972, **52** (Nov.)
'Mersey Kingsway tunnel: planning and design',
  Megaw, T. M. and Brown, C. D.,
'Ditto: construction',
  McKenzie, J. C. and Dodds, G. S.,
'Ditto: project organization, approach roads, buildings and operating organization',
  Cairncross, A. A. and Jones, S. T., *P.*, 1972, **51** (Mar.), disc. 1973, **54** (May)
'Design aspects of tunnelling in Tyneside',
  Boden, J. B., *P.*, 1972, **53** (Sep.)
'Construction of the inclined penstocks at Churchill Falls power station, Labrador',
  Dawson, H. J. and Lyttle, R. P. T., *P.*, 1972, **52** (Nov.), disc. 1973, **54** (Aug.)
'Shaft sinking by ground freezing: Ely Ouse-Essex scheme',
  Collins, S. P. and Deacon, W. G., *P.*, 1972, **Suppl**, disc. 1972, **Suppl**
'The use of rock tunnel machines',
  Pirrie, N. D., *P.*, 1973, **54** (Feb.), disc. 1973, **54** (Nov.)
'Deformation of tunnel with variable ground reaction',
  Brownrigg, D. R. K. and Wood, W. L., *P.*, 1973, **55** (June)
'Tunnel under the Severn and Wye estuaries',
  Haswell, C. K., *P.*, 1973, **54** (Aug.), disc. 1974, **56** (Aug.)
'The bentonite tunnelling machine',
  Bartlett, J. V. and others, *P.*, 1973, **54** (Nov.), disc. 1974, **56** (Aug.)
'The Elbe tunnel, Hamburg: southern ramp and submerged tunnel',
  Havno, K., *P.*, 1974, **56** (Aug.), disc. 1975, **58** (May)
'Soft ground tunnelling',
  Bartlett, J. V. and King, J. R. J., *P.*, 1975, **58** (Nov.), disc. 1976, **60** (Aug.)
'Planning and design of the Hong Kong Mass Transit Railway',
  Edwards, J. T., *P.*, 1976, **60** (Feb.), disc. 1976, **60** (Aug.)
'Development of Edinburgh's sewage disposal scheme',
  Crockett, A. S. and Dugdale, J., *P.*, 1976, **60** (Feb.), disc. 1976, **60** (Nov.)

'Ground movements caused by tunnelling in chalk',
    Priest, S. D., *P.*, 1976, **61** (Mar.), disc. 1976, **61** (Dec.)
'Structural performance of a temporary tunnel lined with spheroidal graphite cast iron',
    Thomas, H. S. H., *P.*, 1976, **61** (Mar.)
'Extension of the Piccadilly Line from Hounslow West to Heathrow Central',
    Jobling, D. G. and Lyons, A. C., *P.*, 1976, **60** (May), disc. 1976, **60** (Nov.)
'Pipe-jacking applied to large structures',
    Clarkson, T. E. and Ropkins, J. W. T., *P.*, 1977, **62** (Nov.), disc. 1978, **64** (Aug.)
'Foyers pumped storage project: planning and design',
    Lander, J. H. and others,
'Ditto: construction',
    Land, D. D. and Hitchings, D. C., *P.*, 1978, **64** (Feb.), disc. 1978, **64** (Nov.)
'Calcutta Rapid Transit System and the Park Street underground station',
    Dasgupta, K. N. and others, *P.*, 1979, **66** (May), disc. 1980, **68** (Feb.)
'Tunnelling trials in chalk',
    O'Reilly, M. P. and others, *P.*, 1979, **67** (June)
'Tyne and Wear metro',
    Howard, D. F., *P.*, 1979, **66** (Aug.)
'The Jubilee Line – 1. The project',
    Cuthbert, E. W.,
'Ditto – 2. Construction from Baker Street to Bond Street exclusive and from Admiralty Arch to Aldwych',
    Lyons, A. C.,
'Ditto – 3. Construction from Bond Street station to Admiralty Arch',
    Bubbers, B. L., *P.*, 1979, **66** (Aug.), disc. 1980, **68** (Feb.)
'Design of concrete shaft linings',
    Auld, F. A., *P.*, 1979, **67** (Sep.)
'The Transclyde railway network: Kelvinhaugh burrowing junction',
    Craige, G. B. and others, *P.*, 1980, **68** (Aug.), disc. 1981, **70** (May)
'Hong Kong Mass Transit Railway Modified Initial System: system planning and multi-contract procedures',
    Edwards, J. T. and others,
'Ditto: design and construction of underground stations and cut-and-cover tunnels',
    McIntosh, D. F. and others,
'Ditto: design and construction of the driven tunnels and the immersed tube',
    Haswell, C. K. and others, *P.*, 1980, **68** (Nov.)
'Tyne and Wear Metro, Byker contract: planning, tunnels, stations and track-work',
    Smyth, W. J. R. and others, *P.*, 1980, **68** (Nov.), disc. 1981, **70** (Aug.)
'Cavitation damage and the Tarbela tunnel collapse of 1974',

Kenn, M. J. and Garrod, A. D., *P.*, 1981, **70** (Feb.), disc. 1981, **70** (Nov.)
'Tyne and Wear Metro: concept, organization and operation',
    Howard, D. F. and Layfield, P.,
'Ditto: management of the project',
    Bartlett, J. V. and others,
'Ditto: design and construction of tunnelling works and underground stations',
    Tough, S. G. and others, *P.*, 1981, **70** (Nov.)

C. AMERICAN SOCIETY OF CIVIL ENGINEERS

From 1867 to 1963 all important papers were published in full, with discussion, in *Transactions* **1–128**; subsequent volumes contain summaries and now only abstracts. Early issues of *Proceedings* were preprints of papers, most of which were subsequently reprinted in *Transactions*. From *Proceedings*, 1956, **82** onwards a series of *Journals of Divisions* has grown, each with a variable number of issues per annum; relevant *Journals* are abbreviated as follows:

Construction CO	*Power PO*
Engineering Issues EI	*Structural ST*
Engineering Mechanics EM	*Surveying and Mapping SU*
City Planning, later *Urban Planning CP/UP*	
Pipe Line, merged in *Transportation Engineering PL/TE*	
Soil Mechanics and Foundations, later *Geotechnical Engineering SM/GT*	

C1 Transactions

'Arching Bergen tunnel on Erie railroad',
    Houston, J., 1872, **1**
'Nesquehoning tunnel',
    Steele, J. D., 1872, **1**
'Tunnels of the Pacific railroad',
    Gilliss, J. R., 1872, **1**
'Reconstruction and enlargement of Cork Run tunnel on the Pittsburgh, Cincinnati and St. Louis railway',
    Becker, M. J., 1877, **6**
'The location of the Chimbote tunnel',
    Nichols, O. F., 1880, **9**
'The Hudson River tunnel',
    Spielmann, A. and Brush, C. B., 1880, **9**
'Shaft sinking under difficulties at Dorchester Bay tunnel, Boston, Mass.',
    Stauffer, D. McN., 1881, **10**
'The Hudson River tunnel',
    Smith, W. S., 1882, **11**
'Tunnel surveying on division no. 6, New Croton aqueduct',
    Watkins, F. W., 1890, **23**

'The Philadelphia tunnel of the Baltimore and Ohio railroad, its construction and cost',
  Thayer, W. W., 1892, **26**
'A method of tunnel alignment',
  Dunham, H. F., 1892, **27**
'Lining a waterworks tunnel with concrete',
  Fitzgerald, D., 1894, **31**
'Brafford's Ridge tunnel',
  Staniford, C. W., 1894, **32**
'Tequixquiac tunnel, Valley of Mexico',
  Campbell, A. J. and Abbot, F. W., 1894, **32**
'The Pennsylvania Avenue subway and tunnel, Philadelphia, Pa.',
  Webster, G. S. and Wagner, S. T., 1902, **48**
'Underground railways: the metropolitan system of Paris',
  Biette, L.,
'Underground railways in Great Britain',
  Mott, B. and Hay, D.,
'Underground railways in the United States',
  Parsons, W. B., 1905, **54(F)**
'The Scranton tunnel of the Lackawanna and Wyoming Valley railroad',
  Francis, G. B. and Dennis, W. F., 1906, **56**
'The bracing of trenches and tunnels, with practical formulas for earth pressures',
  Meem, J. C., 1908, **60**
'Caisson disease and its prevention',
  Japp, H., 1909, **65**
'The New York Tunnel extension of the Pennsylvania railroad: Symposium of 16 papers',
  Raymond, C. W. and others, 1910, **68** and **69**
'Experiments on retaining walls and pressures on tunnels',
  Cain, W., 1911, **72**
'Sinking a wet shaft',
  Hogan, J. P., 1911, **73**
'The Detroit River Tunnel',
  Kinnear, W. S., 1911, **74**
'Air resistance to trains in tube tunnels',
  Davies, J. V., 1912, **75**
'Notes on a tunnel survey',
  Noble, F. C., 1912, **75**
'Specifications for the design of bridges and subways',
  Seaman, H. B., 1912, **75**
'The Laramie-Poudre tunnel',
  Coy, B. G., 1912, **75**

'The Sixth Avenue subway of the Hudson and Manhattan railroad',
    Burrowes, H. G., 1913, **76**
'Subaqueous highway tunnels',
    Snyder, G. D., 1915, **78**
'Secure subway supports',
    Lueder, A. B., 1916, **80**
'The Astoria tunnel under the East River for gas distribution in New York City',
    Davies, J. V., 1916, **80**
'Construction methods for Rogers Pass tunnel',
    Dennis, A. V., 1917, **81**
'Tunnel work on sections 8, 9, 10 and 11, Broadway–Lexington Avenue subway, New York City',
    Werbin, I. V., 1917, **81**
'Underpinning Trinity Vestry building for subway construction',
    Parsons, H. de B., 1917, **81**
'Construction problems of the Manhattan–Bronx and Lexington Avenue subway junction and Queensborough tunnel connections',
    Perrine, G., 1918, **82**
'Notes on tunnel lining for soft ground',
    Johannesson, S. and Hewett, B. H. M., 1920, **83**
'Grouting operations, Catskill water supply',
    Sanborn, J. F. and Zipser, M. E., 1920, **83**
'The proposed New York and New Jersey vehicular tunnel',
    Byrne, E. A. and others, 1920, **83**
'Bridge versus tunnel for the proposed Hudson River crossing at New York City',
    Waddell, J. A., 1921, **84**
'Construction progress of the Hetch Hetchy water supply of San Francisco',
    O'Shaughnessy, M. M., 1922, **85**
'Elastic stresses in the rock surrounding pressure tunnels',
    Dunn, C. P., 1923, **86**
'Engineering geology of the Catskill water supply',
    Berkey, C. P. and Sanborn, J. F., 1923, **86**
'Pipe tunnel under Gowanus canal, Brooklyn, New York',
    Stiles, L. S., 1927, **90**
'Construction methods on the Moffat tunnel',
    Keays, R. H., 1928, **92**
'The Shandakan tunnel',
    Gaussmann, R. W., 1928, **92**
'Methods used in the construction of twelve pre-cast concrete segments for the Alameda County, California, estuary subway',
    Horwege, A. A., 1929, **93**
'Trans-mountain water diversions: a Symposium',
    Debler, E. B. and others, 1930, **94**

'Completion of the Moffat tunnel of Colorado',
  Betts, C. A., 1931, **95**
'The eight-mile Cascade tunnel, Great Northern railway: a Symposium',
  Kerr, D. J. and others, 1932, **96**
'Fulton Street, East River tunnels, New York',
  Killmer, M. I., 1933, **98**
'Analysis of stresses in subaqueous tunnel tubes',
  Eremin, A. A., 1938, **103**
'Stress distribution around a tunnel',
  Mindlin, R. D., 1940, **105**
'Bridge and tunnel approaches',
  Curtin, J. F., 1941, **106**
'Measuring the potential traffic of a proposed vehicular crossing',
  Cherniack, N., 1941, **106**
'Tunnel construction, Sixth Avenue subway, New York',
  Feld, J., 1942, **107**
'Liner-plate tunnels on the Chicago subway',
  Terzaghi, K., 1943, **108**
'Earth pressure on tunnels',
  Housel, W. S., 1943, **108**
'The Queen's Midtown tunnel',
  Singstad, O., 1944, **109**
'Pressure on the lining of circular tunnels in plastic soils',
  Krynine, D. P., 1945, **110**
'Stresses in the linings of shield-driven tunnels',
  Bull, A., 1946, **111**
'Application of geology to tunneling problems',
  Wahlstrom, E. E., 1948, **113**
'Friction coefficients in a large tunnel',
  Peterka, A. J. and Elder, R. A., 1948, **113**
'Diversion tunnel and power conduit of Nantahala hydroelectric development',
  Bleifuss, D. J., 1951, **116**
'Floating tunnel for long water crossings',
  Andrew, C. E., 1951, **116**
'Construction of the Elizabeth River tunnel',
  Peraino, J., 1955, **120**
'Friction measurements in the Apalachia tunnel',
  Elder, R. A., 1958, **123**
'Haas hydroelectric power project',
  Cooke, J. B., 1959, **124**

C2 Proceedings
'Deas Island Tunnel',
  Hall, P., 1957, **83** (ST6, Nov.)

'Geological factors in tunnel construction: 3 papers',
    Cleaves, A. B. and others, 1958, **84** (SM2, May)
'Evaluation of alternative subway routes',
    Griffith, B. A. and von Cube, H. G., 1960, **86** (CP1, May)
'Soil freezing to reconstruct a railway tunnel',
    Low, G. J., 1960, **86** (CO3, Nov.)
'Oahe dam: influence of shale on Oahe power structures design',
    Johns, E. A. and Burnett, R. G., 1963, **89** (SM1, Feb.)
'Design and performance of Mammoth Pool power tunnel',
    Laverty, B. R. and Ludwig, K. R., 1963, **89** (PO1, Sep.)
'Stresses at tunnel intersections',
    Riley, W. F., 1964, **90** (EM2, Apr.)
'Subway tunnel construction in New York City',
    Knight, G. B., 1964, **90** (CO2, Sep.)
'Tunnels for hydroelectric power in Tasmania',
    Thomas, H. H. and Whitman, 1964, **90** (PO3, Oct.)
'Rock trap experience in unlined tunnels',
    Mattimoe, J. J. and others, 1964, **90** (PO3, Oct.)
'Unlined tunnels of the Snowy Mountains Hydroelectric Authority, Australia',
    Dann, H. E. and others, 1964, **90** (PO3, Oct.)
'Contractor's view on unlined tunnels',
    Petrofsky, A. M., 1964, **90** (PO3, Oct.)
'Unlined tunnels of the Southern California Edison Company',
    Spencer, R. W. and others, 1964, **90** (PO3, Oct.)
'Machine tunneling on Missouri River dams',
    Underwood, L. B., 1965, **91** (CO1, May)
'British grouting practice on granular soils',
    Perrott, W. E., 1965, **91** (SM6, Nov.)
'Tunnel design in massive rocks',
    Adler, L., 1966, **92** (SM6, Nov.)
'Stability of clay at vertical openings',
    Broms, B. T. and Bennermark, H., 1967, **93** (SM1, Jan.)
'Oroville dam diversion tunnels',
    Lanning, C. C., 1967, **93** (PO2, Oct.)
'Homestake trans-divide water supply project',
    Thon, G. J. and Campbell, D. E., 1967, **93** (PL3, Nov.)
'Ground freezing in construction',
    Sanger, F. J. and Golder, H. Q., 1968, **94** (SM1, Jan.)
'Stresses against underground structural cylinders',
    Hoeg, K., 1968, **94** (SM4, July)
'What's ahead for tunneling machines',
    Hill, G., 1968, **94** (CO2, Oct.)

'Towards an optimum solution for underground rapid transit',
    Caspe, M. S., 1969, **95** (TE1, Feb.)
'Overall planning of the Bay Area Rapid Transit project',
    Hammond, D. G., 1969, **95** (TE2, May)
'Analysis of tunnel liner-packing systems',
    Dawkins, W. P., 1969, **95** (EM3, June)
'Earthquake design criteria for subways',
    Kuesal, T. R., 1969, **95** (ST6, June)
'Gravity and gradients of long tunnels',
    Williams, H. S., 1969, **95** (SU1, Oct.)
'Chemical grouting for Paris rapid transit tunnels',
    Janin, J. J. and Le Sciellour, G. F., 1970, **96** (CO1, June)
'River diversion for Boundary dam',
    Schilling, A. A., 1970, **96** (PO3, June)
'Diversion at Portage Mountain project',
    Pratt, H. K., 1970, **96** (PO3, June)
'Rock properties and steel tunnel liners',
    Kruse, G. H., 1970, **96** (PO3, June)
'Planning for the Metro',
    Deen, T. B., 1970, **96** (TE3, Aug.)
'Laser alignment techniques utilized in tunneling',
    Cooney, A., 1970, **96** (SU2, Sep.)
'Development of tunneling methods and controls',
    Armstrong, E. L., 1970, **96** (CO2, Oct.)
'Cabin Creek pumped storage hydroelectric project',
    Hight, H. W., 1971, **97** (PO1, Jan.)
'Flathead railroad tunnel construction',
    Sacrison, H., 1971, **97** (CO1, Mar.)
'Design of seismic joint for San Francisco Bay tunnel',
    Douglas, W. S. and Warshaw, R., 1971, **97** (ST4, Apr.)
'Fourth dimension for urban environment',
    Sorensen, K. E., 1971, **97** (UP1, Apr.)
'Surveys for South Shore rapid transit construction',
    Pacelli, A. J., 1971, **97** (SU1, May)
'Precast concrete tunnel linings for Toronto subway',
    Bartlett, J. V. and others, 1971, **97** (CO2, Nov.)
'Sixty years of rock tunneling in Chicago',
    Pikarsky, M., 1971, **97** (CO2, Nov.)
'Vehicular tunnels in rock, directions for development',
    Robbins, R. J., 1972, **98** (CO2, Sep.)
'Computer simulation of subway environment',
    Hoover, T. E. and others, 1973, **99** (TE1, Feb.)

'Recent development of New Austrian Tunneling Method',
    Nussbaum, H., 1973, **99** (CO1, July)
'Tunnel survey and tunneling machine control',
    Peterson, E. W. and Frobenius, P., 1973, **99** (SU1, Sep.)
'Escalators in rapid transit stations',
    O'Neill, R. S., 1974, **100** TE1, Feb.)
'BART in operation – innovations in rapid transit',
    Bugge, W. A., 1974, **100** (TE2, May)
'Subsurface construction contracts, a contractor's view',
    Fox, G. A., 1974, **100** (CO2, June)
'Planning economical tunnel surveys',
    Thompson, B. J., 1974, **100** (SU2, Nov.)
'Subsidence over soft ground tunnel',
    Butler, R. A. and Hampton, D., 1975, **100** (GT1, Jan.)
'Barjansky's shallow-tunnel solution',
    Agrawal, J. S. and Richards, R., 1975, **101** (EM1, Feb.)
'Developments in trench-type tunnel construction',
    Palmer, W. F. and Roberts, K. C., 1975, **101** (CO1, Mar.)
'Liner-medium interaction in tunnels',
    Mohraz, B. and others, 1975, **101** (CO1, Mar.)
'Introduction to fifty years of tunneling',
    Mayo, R. S., 1975, **101** (CO2, June)
'Fifty-year highlights of tunneling equipment',
    Day, D. A. and Boiser, B. P., 1975, **101** (CO2, June)
'Importance of groove spacing in tunnel boring machine operations',
    Rad, P. F., 1975, **101** (GT9, Sep.)
'Karl Terzaghi and the Chicago subway',
    Peck, R. B., 1975, **101** (EI4, Oct.)
'Ventilated approach regions for railway tunnels',
    Vardy, A. E., 1975, **101** (TE4, Nov.)
'Design of transit structures for urban areas',
    Hammond, D. G. and Desai, D. B., 1976, **102** (TE1, Feb.)
'Tunnel and shaft sealing methods using polyvinyl chloride sheet',
    Peduzzi, A., 1976, **102** (CO1, Mar.)
'Analysis of backpacked liners',
    Dar, S. M. and Bates, R. C., 1976, **102** (GT7, July)
'Model Study of tunnel pre-reinforcement',
    Korbin, G. E. and Brekke, T. L., 1976, **102** (GT9, Sep.)
'Applications of photoelasticity to structural analysis',
    Jullien, J. F. and others, 1976, **102** (EM5, Oct.)
'Stresses in shallow elliptical tunnel',
    Mark, R., 1976, **103** (GT11, Nov.)

'Urban tunnels — an option for the transit crisis',
Walton, M. S. and Proctor, R. J., 1976, **102** (TE4, Nov.)
'Half a century progress in soft ground tunneling',
Fox, G. A. and Nicolau, L. E., 1976, **102** (CO4, Dec.)
'Subsidence above shallow tunnels in soft ground',
Atkinson, J. H. and Potts, D. M., 1977, **103** (GT4, Apr.)
'Segmental liner for soft ground tunnels',
Tartaglione, L. C., 1977, **103** (CO2, June)
'Construction methods in tunneling under a railroad',
Soto, M. H., 1977, **103** (CO3, Sep.)
'Fire protection systems on (WMATA) Metro',
Ell, W. M., 1978, **104** (TE1, Jan.)
'Damage to rock tunnels from earthquake shaking',
Dowding, C. H. and Rozan, A., 1978, **104** (GT2, Feb.)
'Model tests for plastic response of lined tunnels',
Kennedy, T. C. and Lindberg, H. E., 1978, **104** (EM2, Apr.)
'Evaluation of alternative station spacings for rapid transit lines',
Permut, H., 1978, **104** (TE3, May)
'Field study of tunnel prereinforcement',
Korbin, G. E. and Brekke, T. L., 1978, **104** (GT8, Aug.)
'Case study of Buffalo's rail transit development',
Wilson, T. M., 1978, **104** (TE5, Sep.)
'Application of epoxy resins in tunnel lining concrete repair',
Borden, R. C. and Selander, C. E., 1978, **98** (CO2, Sep.)
'Temporary tunnel support by artificial ground freezing',
Jones, J. S. and Brown, R. E., 1978, **104** (GT10, Oct.)
'Simplified analysis for tunnel supports',
Einstein, H. H. and Schwartz, C. W., 1979, **105** (GT4, Apr.)
'Pipe jacking state-of-the-art',
Drennon, C. B., 1979, **105** (CO3, Sep.)
'Settlements of immersed tunnels',
Schmidt, B. and Grantz, W. C., 1979, **105** (GT9, Sep.)
'Park River auxiliary tunnel',
Blackey, E. A., 1979, **105** (CO4, Dec.)
'The Central Artery project, construction techniques',
Sholock, F. R., 1980, **106** (TE2, Mar.)
'Economic potential of tunnel standardisation',
Hampton, D. and McCusker, T. G., 1980, **106** (CO3, Sep.)
'Mixed face tunneling on Melbourne underground',
Petrofsky, A. M., 1980, **106** (CO3, Sep.)
'Ground control for shallow tunnels by soil grouting',
Tan, D. Y. and Clough, G. W., 1980, **106** (GT9, Sep.)

'Remote sensing for tunnel siting studies',
    Russell, O. R. and others, 1980, **106** (TE5, Sep.)
'Tokyo's Dainikoro underwater tube tunnel',
    Paulson, B. S., 1980, **106** (CO4, Dec.)

D. OTHER ORGANISATIONS

Four organisations having a substantial and continuing output of material directly relevant to tunnelling are listed separately:

D1 British Tunnelling Society (BTS)
D2 International Tunnelling Association (ITA)
D3 Construction Industry Research and Information Association (CIRIA)
D4 Transport and Road Research Laboratory (TRRL)

There are a growing number of other organisations, including many national societies affiliated to ITA, holding conferences and publishing reports and journals, some of which are mentioned in Sections A, E and F.

The **International Union of Public Transport (UITP)** primarily represents operators, and their quarterly *Revue* and bienneial *Congresses* contain many references to Metros. UITP also publish a quarterly *Biblio-Index* and occasional *Metro Bibliography*.

In the United States of America, the **U.S. National Committee on Tunneling Technology** issue their quarterly *Tunneling Technology Newsletter*; the **American Underground Space Association** have *Underground Space* (Pergamon Press, 6 issues per annum); while there is a steady flow of *Research Reports* from universities, consulting engineers and contractors sponsored by the various administrations in the **U.S. Department of Transportation**.

In the United Kingdom, besides the learned Institutions devoted to Engineering, Geology, Metallurgy and Mining, there are societies for enthusiasts such as the **London Underground Railway Society, Tunnel Study Society** and **Subterranea Britannica**.

In an English language bibliography there is no space to give adequate coverage to **STUVA (Studiengesellschaft fur unterirdische Verkehrsanlagen)** or West Germany whose volumes of *Forschung + Praxis U-Verkehr und Unterirdisches Bauen* have made a valuable contribution.

D1 British Tunnelling Society

This was established in 1971, with the aid of the Institution of Civil Engineers, and holds frequent meetings for presentation of papers and discussions. Its meetings have been reported in some detail in the journal *Tunnels and Tunnelling* and the following selected references are to subjects of general and continuing interest. The dates are those of publication.
'Standardisation in tunnelling',
    1971, **3** (July)

'Efficient excavation (hard rock machine heads)',
 1972, **4** (Jan., Mar., May)
'Ventilation of road tunnels',
 1972, **4** (May, July, Sep.)
'Contractual risks in tunnelling — how should they be shared?',
 1973, **5** (Nov.), 1974, **6** (Mar.)
'Expanded tunnel linings',
 1974, **6** (Mar., May)
'Compressed air and the new decompression table',
 1974, **6** (Nov.), 1975, **7** (Jan.)
'Geological information for underground works',
 1975, **7** (Jan., Mar.)
'Extending the Paris Metro',
 1975, **7** (May, July)
'Problems of shaft sinking',
 1975, **7** (July, Sep.)
'Instrumentation and monitoring of tunnels',
 1975, **7** (Sep., Nov.)
'Problems of tunnelling in chalk',
 1975, **7** (Nov.), 1976, **8** (Jan.)
'Early history of sapper tunnelling',
 1976, **8** (Jan., Mar.)
'Blasting in urban areas',
 1976, **8** (Sep.), 1977, **9** (Mar.)
'Probing ahead for tunnels',
 1977, **9** (Jan.)
'Tunnelling trials in chalk',
 1977, **9** (Mar.)
'Tunnelling — improved contract practices',
 1977, **9** (Nov.)
'U.S. and U.K. tunnelling — developments, comparisons and trends',
 1978, **10** (Jan.)
'Lining and waterproofing techniques in Germany',
 1978, **10** (Apr., May)
'Hydraulics versus pneumatics',
 1978, **10** (July)
'Modern surveying instruments',
 1978, **10** (Dec.)
'Immersed tunnels — Danish style',
 1979, **11** (July)
'Urban tunnelling in the eastern U.S.',
 1979, **11** (July)

'Ground classification – continental and British',
   1980, **12** (July)

D2 International Tunnelling Association – ITA
(Association Internationale des Travaux en Souterrain – AITES, 109 Av. Salvador
Allende, 69500 Bron, France), was formed in 1974 as an international commission on tunnelling activities. Almost all major countries with any interest in
tunnels are represented by their national Society, Institution or Authority. In
recent years the most significant functions of ITA have tended to move away
from its international congresses into the activities of nine working groups on
the following subjects briefly described in *Tunnels and Tunnelling*:

Contractual sharing of risks	1980, **12** (Jan)
Standardization	(Mar)
Research	(Apr)
Safety in works	(May)
Catalogue of works in progress	(June)
Subsurface planning	(Oct)
Structural design models	(Nov)
Seismic effects	(Dec)
Maintenance and repair	1981, **13** (Jan)

The *Proceedings* of the first four *Annual Meetings* and the first three *Reports by
the Working Groups* were published by AITES, Bron. All subsequent *Proceedings*
and *Reports* and published in *Advances in Tunnelling Technology and Subsurface Use*, from 1980, **1** (1), Pergamon Press, 3 issues per annum.

D3 Construction Industry Research and Information Association (CIRIA)
Founded in 1964 as the Civil Engineering Research Association (CERA), CIRIA
is an association of contractors, manufacturers, consulting engineers, universities,
local and public authorities. A sectional committee on Underground Construction
was formed in 1974. Results of research projects are published as *Reports* (*Rep*)
and *Technical Notes* (*TN*).
'Hydraulic design of unlined and lined-invert rock tunnels',
   Wright, D. E., *Rep*. 29, 1971
'Full-scale tests to determine the hydraulic performance of a lined-invert rock
tunnel',
   Wright, D. E., *TN* 16, 1971
'Medical Code of Practice for work in compressed air',
   Rep. 44, 1973, 2nd edn. 1975, 3rd edn. 1981
'The economics of vehicular tunnel ventilation and the automatic control of
tunnel systems',
   Pursall, B. R. and Swann, C. D., *TN* 53, 1974
'Experiences with a new decompression table for work in compressed air',
   TN 59, 1974

'Classified index of research requirements in underground construction',
    1975
'Recommended tunnel sizes – a cost benefit study',
    Rep. 66, 1977
'Lateral movement of heavy loads',
    Rep. 68, 1977
'Recording drilling performance for tunnelling site investigations',
    Brown, E. T. and Phillips, H. R., *TN* 81, 1977
'Down hole instrumentation – a review of tunnelling ground investigation',
    TN 90, 1977
'Tunnelling, improved contract practices',
    Rep. 79, 1978
'A review of instruments for gas and dust monitoring underground',
    Rep. 80, 1978
'Tunnel waterproofing',
    Rep. 81, 1979
'Precast concrete tunnel linings – review of current test procedures',
    Lance, G. A., *TN* 104, 1981
'Acoustic techniques for rock quality evaluation',
    TN in prep.
'Guide to the health and safety aspects of ground treatment materials',
    Rep. in prep.
'Hydraulic roughness of segmentally-lined tunnels',
    Rep. in prep.
'Rock reinforcement: a state-of-the-art review',
    Rep. in prep.

D4 Transport and Road Research Laboratory (TRRL)

Founded in 1933 as the Road Research Laboratory (RRL), the TRRL is currently
responsible to the Departments of the Environment and Transport. In 1970
there was formed what has now become the Tunnels and Underground Pipes
Division. Results of research are published as either *Laboratory Reports* (*LR*) or
Supplementary Reports (*SR*).
'The classification and description of soils for engineering purposes: a suggested
revision of the British system',
    Dumbleton, M. J., *LR* 182, 1968
'The effect of tides on subsoil pore-water pressures at a site on the proposed
Shoreham by-pass',
    Margason, G. and others, *LR* 195, 1968
'A preliminary study of the cost of tunnel construction',
    Margason, G. and Pocock, R. G., *LR* 326, 1970
'Preliminary sources of information for site investigations in Britain',
    Dumbleton, M. J. and West, G., *LR* 403, 1971, revised 1976

'Channel Tunnel: vehicular movements in proposed ferry trains',
  Ellson, P. B. and Layfield, R. E., *LR* 435, 1972
'Channel Tunnel: pedestrian movements in proposed ferry trains',
  Cundill, M. A., *LR* 436, 1972
'A review of methods of ventilating road tunnels',
  Hignett, H. J. and Hogbin, L., *LR* 477, 1972
'An analysis of vehicle breakdowns in the Mersey Tunnel',
  Bartlett, R. S., and Chhotu, S. R., *LR* 484, 1972
'Some examples of underground development in Europe',
  O'Reilly, M. P., *LR* 592, 1974
'Guidance on planning, directing and reporting site investigations',
  Dumbleton, M. J. and West, G., *LR* 625, 1974
'A pilot scale machine for tunnel boring research',
  Hignett, H. J. and Howard, T. R., *LR* 632, 1974
'Measurement of ground movements during a bentonite tunnelling experiment',
  Boden, J. B. and McCaul, C., *LR* 653, 1974
'Techniques and equipment for detecting underground services',
  Keir, W. G., *SR* 69, 1974
'Note on a visit to Italy – September 1973',
  O'Reilly, M. P., *SR* 80, 1974
'A horizontal inclinometer for measuring ground movements',
  Hudson, J. A. and Morgan, J. M., *SR* 92, 1974
'An assessment of geophysics in site investigation for roads in Britain',
  West, G. and Dumbleton, M. J., *LR* 680, 1975
'The influence of tunnel excavation on an adjacent shaft in chalk',
  Hudson, J. A. and McCaul, C., *SR* 161, 1975
'Probing ahead for tunnels: a review of present methods and recommendations
for research',
  SR 171, 1975
'Rock quality in the Kielder Experimental Tunnel, Co. Durham',
  Priest, S. D. and Hudson, J. A., *SR* 173, 1975
'Research studies into the behaviour of tunnels and tunnel linings in soft ground',
  Atkinson, J. H. and others, *SR* 176, 1975
'Measurements of ground movement and lining behaviour on the London Under-
ground at Regents Park',
  Barratt, D. A. and Tyler, R. G., *LR* 684, 1976
'An impact penetrometer for assessing the cuttability of soft rocks',
  Hudson, J. A. and Drew, S. D., *LR* 685, 1976
'The particle-size distributions of debris produced during tunnelling trials',
  Toombs, A. F. and others, *LR* 714, 1976
'Ultrasonic wave propagation in discontinuous rock',
  New, B. M., *LR* 720, 1976

'Site investigation for tunnelling trials in chalk',
  Priest, S. D., *LR* 730, 1976
'A guide to site investigation procedures for tunnels',
  Dumbleton, M. J. and West, G., *LR* 740, 1976
'Noise measurements during tunnelling trials at Chinnor',
  Foster, K. and others, *SR* 188, 1976
'Machine performance data recorded during the Chinnor tunnelling trials',
  Snowdon, R. A., *SR* 196, 1976
'Measurement of ground movement due to excavation of a shallow tunnel in lower chalk',
  McCaul, C. and others, *SR* 199, 1976
'A note on a use of stereophotography in tunnelling',
  West, G., *SR* 211, 1976
'Shotcreting on chalk: some preliminary trials',
  McCaul, C. and Ryley, M. D., *SR* 212, 1976
'Traffic induced ground vibration in the vicinity of road tunnels',
  Bean, R. and Page, J., *SR* 218, 1976
'Site investigation aspects of the Sydenham Road sewer tunnel',
  Dumbleton, M. J. and Toombs, A. F., *SR* 235, 1976
'The tunnelling system for the British section of the Channel tunnel phase II works',
  Morgan, J. M. and others, *LR* 734, 1977
'Tunnelling trials in chalk: rock cutting experiments',
  Hignett, H. J. and others, *LR* 796, 1977
'Site investigation aspects of the River Tyne Siphon Sewer Tunnel',
  Dumbleton, M. J. and Priest, S. D., *LR* 831, 1978
'Site investigation aspects of the Empingham reservoir tunnels',
  Dumbleton, M. J. and Toombs, A. F., *LR* 845, 1978
'The effects of ground vibration during bentonite shield tunnelling at Warrington',
  New, B. M., *LR* 860, 1978
'A review of tunnel lining methods in the United Kingdom',
  Craig, R. N. and Muir Wood, A. M., *SR* 335, 1978
'The arrangements of rock cutting tools on full face tunnel boring machines',
  Hignett, H. J. and O'Reilly, M. P., *SR* 376, 1978
'Settlements caused by tunnelling in weak ground at Stockton-on-Tees',
  McCaul, C., *SR* 383, 1978
'The propagation of ultrasonic waves in strong rocks',
  Toombs, A. F., *SR* 384, 1978
'Site investigation aspects of the River Medway cable tunnels',
  Dumbleton, M. J. and others, *SR* 451, 1978
'Site investigation aspects of part of the Tyneside South Bank interceptor sewer',
  Dumbleton, M. J. and others, *SR* 454, 1978

'Site investigation and construction of the Liverpool Loop and Link Tunnels',
    West, G. and Toombs, A. F., *LR* 868, 1979
'Speed/flow relationships in road tunnels',
    Bampfylde, A. P. and others, *SR* 455, 1979
'Tunnel boring machine performance and ground properties. Report on the initial 1½ km of the North Wear Drive, Kielder Aqueduct',
    Morgan, J. M. and others, *SR* 469, 1979
'A preliminary study of the reproducibility of joint measurements in rock',
    West, G., *SR* 488, 1979
'Tunnelling machine instrumentation for use in coal mines',
    Temporal, J. and Hignett, H. J., *SR* 502, 1979
'Geophysical and television borehole logging for probing ahead of tunnels',
    West, G., *LR* 932, 1980
'Measurements of ground movements around three tunnels in loose cohesionless soil',
    Ryley, M. D. and others, *LR* 938, 1980
'Some observations of machine tunnelling at the Kielder Aqueduct',
    O'Rourke, T. D. and others, *SR* 532, 1980
'Settlement caused by tunnelling beneath a motorway embankment',
    Toombs, A. F., *SR* 547, 1980
'Site investigation and construction of the Dinorwic diversion tunnel',
    West, G. and Ewan, V. J., *LR* 984, 1981
'Site investigation and construction of the Cardiff cable tunnel',
    West, G. and McLaren, D., *LR* 1012, 1981
'Reproducibility of joint spacing measurements in rock',
    Ewan, V. J. and others, *LR* 1013, 1981
'Kielder aqueduct tunnels – predicted and actual geology',
    Davis, T. P. and others, *SR* 676, 1981

E. CONFERENCES AND SYMPOSIA

The world-wide growth of such meetings presents difficulties in keeping track of the information presented. The papers vary in their importance as to permanent value, wide interest, repetition elsewhere, and inclusion or absence of discussion. The section is divided:

E1 Tunnelling Conferences
E2 Other related conferences with a significant tunnelling content.

E1 Tunnelling Conferences

Symposium on shaft sinking and tunnelling, Instn. Min. Engrs, London, 1959
Rapid excavation – problems and progress; Proc. Tunnel and Shaft Conf.,
    A.I.M.E., Minneapolis, 1968
Large permanent underground openings, Proceedings of the International Symposium, Intnl Soc. Rock Mechs, Oslo, 1969

Deep tunnels in hard rock: a solution to combined sewer overflow and flooding problems, U.S. Environmental Protection Agency, Milwaukee, 1970

The technology and potential of tunnelling: South African Tunnelling Conference, S. African Instn Civ. Engrs, Johannesburg, 1970

6th National Tunnel Symposium, Japan Soc. Civ. Engrs, Tokyo, 1970

Advisory Conference on Tunnelling, Organisation for Economic Co-operation and Development, Washington, 1970

Seminar on Engineering problems in tunnelling, Indian Soc. Engng Geol., New Delhi, 1971

Symposium on Underground Rock Chambers, Amer. Soc. Civ. Engrs, Phoenix, 1971

Proceedings, International Symposium on Underground openings, Swiss Soc. Soil Mechs Fndtn Engng, Lucerne, 1972

Proceedings of Rapid Excavation and Tunneling Conferences, 1st, Chicago, 1972; 2nd, San Francisco, 1974; 3rd, Las Vegas, 1976; 4th, Atlanta, 1979; 5th, San Francisco, 1981, Society of Mining Engineers of A.I.M.E.

Tunnelling in Rock, a course of lectures, S. African Instn Civ. Engrs, Scientia, 1973

Proceedings of a symposium on rock mechanics in tunnelling problems, Indian Geotechnical Society, Haryana, 1973

International Symposium on the Aerodynamics and Ventilation of Vehicle Tunnels, BHRA Fluid Engineering, 1st, Canterbury, 1973; 2nd, Cambridge, 1976; 3rd, Sheffield, 1979

Use of shotcrete for underground structural support, Amer. Soc. Civ. Engrs, South Berwick, 1973

Shotcrete for ground support, Amer. Soc. Civ. Engrs, Easton, 1976

Shotcrete for underground support III, Engineering Foundation, St. Anton am Arlberg, 1978

Subsurface exploration for underground excavation and heavy construction, Amer. Soc. Civ. Engrs, Henniker, 1974

Australian Tunnelling Conferences, 1st, *Reshaping cities using underground construction*, Melbourne, 1974; 2nd, *Design and construction of tunnels and shafts*, Melbourne, 1976; 3rd, Sydney, 1978; 4th, Melbourne, 1981, Australian Tunnelling Association

Hazards in tunnelling and on falsework; Proc. 3rd Intnl Safety Conf., Instn Civ. Engrs, London, 1975

Mechanical boring or drill and blast tunnelling: 1st U.S.–Swedish underground workshop, Swedish Detonic Research Foundation, Stockholm, 1976

Tunnelling '76: Proceedings of an international Symposium, Instn. Min. Metal., London, 1976

Tunnelling '79: Proceedings of the 2nd international Sympsoium, Instn Min. Metal., London, 1979

Computer methods in tunnel design, Instn Civ. Engrs, London, 1977

Storage in excavated rock caverns, Rockstore '77, Proc. 1st Intnl Symp., Stockholm, 1977 (publ. Pergamon)

Subsurface Space, Rockstore '80, Proc. Intnl Symp., Stockholm, 1980 (publ. Pergamon)

Immersed Tunnels: Delta Tunnelling Symposium, Royal Instn Engrs, Netherlands, Amsterdam, 1978

International Symposium on tunnelling under difficult conditions, Japan Tunnelling Assoc., Tokyo, 1978

Seminar on utility services underground, Royal Society of Arts, London, 1978

Eurotunnel Conference, Basle, 1978 and 1980

Analysis of tunnel stability by the convergence-confinement method, Proc. Conference, Paris, 1978; in *Underground Space*, 1980, **4** (4, 5 & 6)

International Symposium, the safety of underground works, Brussels, 1980

E2 Other Conferences

PIARC World Road Conferences:

    Road Tunnel Committee *Reports*: 11th, Rio de Janeiro, 1959; 12th, Rome, 1964; 13th, Tokyo, 1967; 14th, Prague, 1971; 15th, Mexico City, 1975; 16th, Vienna, 1979

    Road Tunnel Committee *Documentation Digests*: 13th, Tokyo, 1967; 14th, Prague, 1971; 15th, Mexico City, 1975

    Permanent International Association of Road Congresses

Grouts and drilling muds in engineering practice, British National Society of ISSMFE, London, 1963

In situ investigations in soils and rocks, Proceedings of the Conference, British Geotechnical Society, London, 1969

Conferences on Soil Mechanics and Foundation Engineering:

    7th International, Mexico, 1969, esp. **State of the Art**, Peck, R. B., 'Deep excavations and tunneling in soft ground', **2** Session 4, **3** General report and Session 4

    5th Panamerican, Buenos Aires, 1975, esp. **4**, Cording, E. J. and Hansmire, W. H., 'Displacements around soft ground tunnels'

    9th International, Tokyo, 1977, esp. *Speciality Session no. 1*, and **Case History** (both published separately)

    6th Asian, Singapore, 1979, esp. **2**, Watanabe, T., 'Subway tunnelling in soft ground'

    6th Panamerican, Lima, 1979, esp. *Speciality Session on Soft ground tunnelling* (published separately, Balkema, Rotterdam, 1981)

    10th International, Stockholm, 1981, esp. **1** Session 2, and **4** Session 2

Dynamic Rock Mechanics, 12th Symposium on Rock Mechanics, Society of Mining Engineers, AIME, Rolla, Missouri, 1970, esp. Part VI

2nd International Congress of the International Association of Engineering Geologists, Sao Paulo, Brazil, 1974, esp. 2 Theme VII

3rd International Congress of Engineering Geology, Madrid, 1978, esp. 2, Section III-c

Design Methods in Rock Mechanics, 16th Symposium on Rock Mechanics, Amer. Soc. Civ. Engrs, Minneapolis, 1975, esp. Session 3

Proceedings of the Symposium on Exploration for Rock Engineering, S. African Instn Civ. Engrs, Johannesburg, 1976

Proceedings of a Conference on Rock Engineering, British Geotechnical Society, Newcastle-upon-Tyne, 1977

Field Measurements in Rock Mechanics, Proceedings of the International Symposium, ISETH, Zurich, 1977

Large Ground Movements and Structures, Proceedings of the Conference (esp. Session IV and State of the Art), UWIST, Cardiff, 1977

2nd Conference on Ground Movements and Structures (esp. Sessions IV and VII), UWIST, Cardiff, 1980

International Symposium on Ground Freezing, Bochum, W. Germany, 1978

2nd International Symposium on Ground Freezing, Trondheim, Norway, 1980

Mass Transportation in Asia, Concrete Society (HK) Ltd., Hong Kong, 1980

F. JOURNALS AND PERIODICALS

F1 contains a selection of papers from *Tunnels and Tunnelling*. First issued in May 1969, now a monthly, it is the premier journal on the subject and the official journal of the British Tunnelling Society; issues reporting some BTS meetings have already been listed in Section D1.

F2 contains a selection of papers from the vast range of engineering and related journals and from a few non-tunnelling conferences.

Apart from *Tunnels and Tunnelling, Tunneling Technology Newsletter* and *Underground Space* (referred to in Section D), there are many periodicals currently carrying up to date news of tunnelling projects, among the prominent are:

Civil Engineering – London
New Civil Engineer
New Civil Engineer International
International Water Power and Dam Construction
Civil Engineering – New York
Engineering News-Record

F1 *Tunnels and Tunnelling*
Selective bibliography on immersed tubes,
    Pequignot, C. A., 1969, **1** (July)
Some aspects of rail tunnel maintenance (UK),
    Walton, J. A., 1969, **1** (July)

Tunnel construction and maintenance on the Austrian Federal Railway system,
  Ziermann, R., 1970, 2 (Mar., May)
Maintaining the tunnels of French Railways,
  Bogaert, G. van den, 1970, 2 (July, Sep.)
Tunnels of Norwegian State Railways,
  Hartmark, H., 1970, 2 (Nov.)
Design, construction and maintenance of tunnels of German State Railways,
  Spang, J., 1971, 3 (Sep.)
Rail tunnels in Czechoslovakia,
  Streit, J., 1973, 5 (May)
Air pollution in vehicular road tunnels,
  Pursall, B. R. and Swann, C. D., 1972, 4 (July)
Some aspects of fire behaviour in tunnels,
  Roberts, A. G., 1973, 5 (Jan.)
Foundation of a tunnel by the sandflow system,
  Griffioen, A. and Van der Veen, R., 1973, 5 (July)
The generation and alleviation of air pressure transients in tunnels,
  Fox, J. A. and Vardy, A. E., 1973, 5 (Nov.)
Subterrene rock melting devices,
  Altseimer, J. H., 1974, 6 (Jan.)
Channel Tunnel survey,
  Hulme, T. W., 1974, 6 (Sep.)
Visco-elastic tunnel analysis,
  Curtis, D. J., 1974, 6 (Nov.)
The Shin Kanmon connection,
  Brown, R. L., 1975, 7 (Jan.)
Immersed-tube tunnels,
  Culverwell, D. R., 1976, 8 (Jan., Mar.)
Mechanised tunnelling – progress and expectations,
  Robbins, R. J., 1976, 8 (May)
A view of the Australian tunnelling scene,
  Neyland, A. J., 1976, 8 (July)
Survey of hydraulic drilling performance,
  Bullock, R. L., 1976, 8 (Sep.)
Road tunnels in Norway,
  Grønhaug, A., 1976, 8 (Nov.)
Pipe jacking case histories,
  1977, 9 (July, Nov.); 1978, 10 (Jan., Mar., Apr., May, June, July)
Some road and rail tunnels in Japan,
  Leeney, J. G., 1977, 9 (Nov.)
Japanese slurry shield tunnelling,
  Leeney, J. G., 1978, 10 (Jan.)

Removing misconceptions on the New Austrian Tunnelling Method,
  Müller, L., 1978, **10** (Oct.)
Another view of the NATM,
  Golser, J., 1979, **11** (Mar.)
The German tunnelling industry – special report,
  Harding, P. G., 1979, **11** (Apr.)
Ground behaviour and support for mining and tunnelling,
  Muir Wood, A. M., 1979, **11** (May, June)
Assessment of tunnel boring machine performance,
  McFeat-Smith, I. and Tarkoy, P. J., 1979, **11** (Dec.)
Tunnel boring machines in difficult ground,
  McFeat-Smith, I. and Tarkoy, P. J., 1980, **12** (Jan.)
Site investigations for machine tunnelling contracts,
  McFeat-Smith, I. and Tarkoy, P. J., 1980, **12** (Mar.)
A history of the Channel tunnel,
  Muir Wood, A. M., 1980, **12** (May)
"Earthworm" system will threaten conventional tunnel jacking,
  Richardson, M. and Scruby, J., 1981, **13** (Apr.)
The Unitunnel in action,
  Richardson, M. and Scruby, J., 1981, **13** (May)
World list of immersed tubes,
  Culverwell, D. R., 1981, **13** (Sep.)
Putting the NATM into perspective,
  Brown, E. T., 1981, **13** (Nov.)

F2 Journals and Periodicals
The Mont Cenis tunnel,
  Kossuth, F., *Engng*, 1871, **11** and **12**, 1872, **13**
The Channel Tunnel,
  Tylden-Wright, C., *Trans. N. England Inst. Min. Mech. Engrs*, 1882, **32**, 3
  and 55
On the construction of the Glasgow City and District railway,
  Simpson, R., *Trans. Instn. Engrs Shipbuilders Scotland*, 1887–88, **31**
Cast-iron segments for railway and other tunnels,
  Carey, E. G.,
Tunnelling in soft materials with special reference to Glasgow District subway,
  Simpson, R., *Trans. Instn Engrs Shipbuilders Scotland*, 1895–96, **39** (Dec.
  24, Jan. 28, Feb. 25)
Construction and demolition of tunnels,
  Ross, M. G., *Trans. Liverpool Engng Soc.*, 1908, **29**
Tunnelling in water-bearing strata,
  Baker, R. F., *Trans. Liverpool Engng Soc.*, 1921, **42**

Tube railway tunnelling,
     Anderson, D., *Trans. Liverpool Engng Soc.*, 1924, **45**
The Holland vehicular tunnel under the Hudson River,
     Skinner, F. W., *Engng*, 1927, **124** (Nov. 11 and 25, Dec. 9)
The Post Office tube railway, London,
     Engng, 1928, **125** (Jan. 27, Feb. 10 and 24, Mar. 2 and 16)
Construction of a tunnel under the River Liffey,
     Nicholls, H., *Trans. Instn Civ. Engrs Ireland,* 1929, **55**
Ventilation of vehicular tunnels,
     Singstad, O., *Proc. World Engng Congr.*, Tokyo, 1929, **9**, 381-99
Driving a mines drainage tunnel in North Wales,
     Francis, J. L. and Allan, J. C., *Trans. Instn Min. Metal.*, 1932, **41**
The ventilation of tunnels,
     Haldane, J. S., *J. Instn Heat. Vent. Engrs*, 1936, **4** (Mar.)
Ventilating the Lincoln vehicular tunnel,
     Murdock, W., *Trans. Amer. Soc. Heat Vent. Engrs*, 1938, **44**, 273-88
Mono Craters tunnel construction problems,
     Jacques, H. L., *J. Amer. Waterworks Assoc.*, 1940, **32**, 43
The road tunnel under the River Maas at Rotterdam,
     Van Bruggen, J. P., *Engng*, 1940, **150** (Aug. 9 and 30, Sep. 27)
A century of tunnelling,
     Halcrow, W. T., *Proc. Instn Mech. Engrs*, 1941, **146**
Piston effect of trains in tunnels,
     Dougherty, R. L., *Trans. Amer. Soc. Mech. Engrs*, 1942, **64** (Feb.)
Shield tunnels of the Chicago subway,
     Terzaghi, K., *J. Boston Soc. Civ. Engrs*, 1942, **29** (July)
Ventilation and cooling in London's tube railways,
     Mount, Sc C., *J. Instn Heat & Vent. Engrs,* 1947, **14** (Jan.-Feb.)
Historical development of subaqueous tunneling,
     Singstad, O., *J. Boston Soc. Civ. Engrs*, 1949, **36**
Underground railway planning and construction,
     Anderson, D., *Proc. Joint Engng Conf.*, London, 1951
The main access tunnel at London Airport,
     Concrete Constr. Engng, 1952, **47** (Mar.)
Stresses in rocks about cavities,
     Terzaghi, K. and Richart, F. E., *Geotechnique*, 1952, **3** (June)
Lining the Bowland tunnel,
     Wat. Pwr., 1953, **5** (Feb.)
Geology of the Bowland Forest tunnel,
     Earp, J. R., *Bull. Geol. Surv. G.B.*, 1955 (7)
Tunnel linings,
     Dawson, O., *Trans. Instn Civ. Engrs Ireland*, 1956, **83**

Modern blasting practice in tunnelling operations,
    Rankin, W. W. and Haslam, R., *Civ. Engng Publ. Wks Rev.*, 1957, **52** (Feb., Mar. and Apr.)

Design and construction of the Hampton Roads tunnel,
    Bickel, J. O., *J. Boston Soc. Civ. Engrs*, 1958, **45** (Oct.)

The construction of the Howth tunnel,
    O.Shee, S. F., *Trans. Instn Civ. Engrs Ireland*, 1958, **84**, 49 and 171

Effect of diesel locomotive operation on atmospheric conditions in a railway tunnel,
    Rennie, R. P. and others, *Can. J. Chem. Engng*, 1960, **38** (Aug.)

The invention and development of injection processes,
    Glossop, R., *Geotechnique*, 1960, **10** (Sep.) and 1961, **11** (Dec.)

A contribution to the analysis of stress in a circular tunnel,
    Morgan, H. D., *Geotechnique*, 1961, **11** (Mar.)

Theory and practice of rock bolting,
    Lang, T. A., *Trans. Soc. Min. Engrs AIME*, 1961, **220**, 333

Technical description of rock cores for engineering purposes,
    Deere, D. U., *Rock Mech. Engng Geol.*, 1963, **1**, 18

The tunnel of Eupalinus (Samos tunnel),
    Goodfield, J., *Scientific American*, 1964, **210** (June)

Temperature and ventilation in the Toronto subway, an experimental study,
    Brown, W. E., *Engr.*, 1964, **218** (Aug. 28)

Christchurch–Lyttleton road tunnel (4 papers),
    Spooner, B. W. and others, *New Zealand Engng*, 1964, **19** (Dec.)

Prestressed concrete in sub-aqueous tunnel construction,
    Hall, P. F., *J. Prestressed Concr. Inst.*, 1965, **10** (Aug.)

Basic theory of rapid-transit tunnel ventilation,
    Brown, W. G., *Trans. Amer. Soc. Mech. Engrs*, 1966, **88B** (Feb.)

Recent developments in the design of submerged tunnels,
    Brink, A., *Struct. Engr.*, 1966, **44** (Feb.); *see also* discussion 1966, **44** (Aug.)

Trench type subaqueous tunnels: design and construction,
    Bickel, J. O., *Struct. Engr.*, 1966, **44** (Oct.); *see also* discussion 1967, **45** (Feb.)

Hard rock tunnelling machines,
    Muirhead, I. R. and Glossop, L. G., *Trans. Instn Min. Metall.*, 1968, **77A**, 1; *see also* discussion, 118

Modern tunnelling methods,
    Donovan, H. J., *Public Works & Municipal Services Congress*, London, 1968

The Gibraltar Hill tunnels, Monmouth (Great Britain),
    Baker, C. O. and Howells, D. A., *Engng Geol. Amsterdam*, 1969, **3**, 121

Seikan undersea tunnel,
    Tanaka, T., *Civil Engineering in Japan*, 1970, **9**

The Tower Subway: the first tube tunnel in the world,
    Lee, C. E., *Trans. Newcomen Soc.*, 1970, **43**
Tunnels for roads and motorways,
    Muir Wood, A. M., *Q. J. Engng Geol.*, 1972, **5** (1 and 2)
Shotcrete support with special reference to Mexico City drainage tunnels,
    Mason, E. E. and Mason, R. E., *Rock Mechs*, 1972, **4**, 115
Seikan undersea tunnel,
    Harma, K., *Japanese Railway Engineering*, 1972 (Jan.)
Structural aspects of submerged tube tunnel construction,
    Williams, G. M. S. and Innes, K. W., *Struct. Engr.*, 1972, **50** (Feb.); *see also*
    discussion 1972, **50** (Aug.)
Long horizontal boring,
    Mocheda, Y., *Permanent Way*, 1973, 14 (Apr.)
Grouting in pilot tunnel-heading,
    Shinohara, H. and Akita, K., *Ibid*
Engineering classification of rock masses for the design of tunnel support,
    Barton, N. and others, *Rock Mechs.*, 1974, **6**, 189
Construction of Shin-Kanmon tunnel,
    Saito, T., *Civ. Engng in Japan*, 1974, **13**
Transient flows in tunnel complexes of the type proposed for the Channel tunnel,
    Henson, D. A. and Fox, J. A., *Proc. Instn Mech. Engrs*, 1974, **188** (15/74)
The design of the Channel tunnel,
    Gould, H. B. and others, *Struct. Engr*, 1975, **53** (Feb.); *see also* discussion,
    1975, **53** (Dec.)
The circular tunnel in elastic ground,
    Muir Wood, A. M., *Geotechnique*, 1975, **25** (Mar.); *see also* discussion, 1976,
    26 (Mar.)
The invention and early use of compressed air to exclude water from shafts and
tunnels during construction,
    Glossop, R., *Geotechnique*, 1976, **26** (June)
Some developments in segmental tunnel linings designed in the United Kingdom,
    Lyons, A. C., *Underground Space*, 1977, **1** (3)
Seikan undersea tunnel,
    Matsuo, S. and Endo, K., *Civil Engineering in Japan*, 1977, **16**
Design of a ventilation system for an English Channel Tunnel,
    Henson, D. A. and Lowndes, J. F. L., *ASHRAE J.*, 1978 (Feb.)
Geotechnical performance of a tunnel in till,
    Eisenstein, Z. and Thomson, S., *Canad. Geotech. J.*, 1978, **15** (Aug.)
The Kaimai Tunnel (9 papers),
    Hermans, R. E. and others, *Trans. New Zealand Instn Engrs*, 1978, **5** (1)
The 50-year history of the subways in Japan,
    Watanabe, T., *Civil Engineering in Japan*, 1978, **17**

Hong Kong MTRC harbour tunnel,
 Hansen, F. J., *Hong Kong Engr*, 1979, **7** (Apr.)
Tunnel design: philosophy and practice,
 Hulme, T. W., *Struct. Engr.*, 1979, **57**A (May); *see also* discussion, 1981,
 59A (Apr.)
The Hong Kong mass transit immersed tube — some aspects of construction,
 Kennedy, M. J., *Hong Kong Engr*, 1980, **8** (Jan.)
Dams and their tunnels,
 Muir Wood, A. M. and others, *Int. Wat. Pwr Dam Constr.*, 1980, **32** (Feb.,
 Mar., Apr. and May)
Field measurements in two tunnels in Edmonton, Alberta,
 Thomson, S. and El-Nahhas, F., *Canad. Geotech. J.*, 1980, **17** (Feb.)
The stability of shallow tunnels and underground openings in cohesive material,
 Davis, E. H. and others, *Geotechnique*, 1980, **30** (Dec.)
An initial appraisal of ground probing radar for site investigation in Britain,
 Darracott, B. W. and Lake, M. I., *Ground Engng*, 1981, **14** (Apr.)
Recent developments in geophysical techniques for the rapid location of near-surface anomalous ground conditions,
 McDowell, P., *Ground Engng*, 1981, **14** (Apr.)
Soil mechanics aspects of soft ground tunnelling,
 Atkinson, J. H. and Mair, R. J., *Ground Engng*, 1981, **14** (July)
Site investigation for tunnels,
 West, G. and others, *Inst. J. Rock Mech. Min. Sci.*, 1981, **18** (5)
The construction of the Ahmed Hamdi road tunnel, Suez,
 Harries, D. A., *Highway Engr.*, 1981, **28** (Nov.)

Index